Sex and the
Origins of Death

Sex and the Origins of Death

William R. Clark

OXFORD UNIVERSITY PRESS

New York   Oxford

Oxford University Press

Oxford   New York
Athens   Auckland   Bangkok   Bogotá   Bombay
Buenos Aires   Calcutta   Cape Town   Dar es Salaam
Delhi   Florence   Hong Kong   Istanbul   Karachi
Kuala Lumpur   Madras   Madrid   Melbourne
Mexico City   Nairobi   Paris   Singapore
Taipei   Tokyo   Toronto   Warsaw

and associated companies in
Berlin   Ibadan

Copyright © 1996 by Oxford University Press, Inc.

Illustrations by Celine Park

First published by Oxford University Press, Inc., 1996

First issued as an Oxford University Press paperback, 1998

Oxford is a registered trademark of Oxford University Press

Library of Congress Cataloging-in-Publication Data
Clark, William R., 1938–
Sex and the origins of death / by William R. Clark.
p.   cm.   ISBN 0-19-510644-X
ISBN 0-19-512119-8 (Pbk.)
1. Cell death. 2. Death (Biology) 3. Apoptosis. 4. Sex (Biology) I. Title.
QH671.C56   1996
574.87'65—dc20   96-11753

10 9 8 7 6 5 4 3 2 1

Printed in the United States of America

Acknowledgments

A number of people have worked hard to help me in various stages of creating this book. At Oxford University Press I have enjoyed the continued encouragement and support of Kirk Jensen and Laura Brown. I am especially indebted to Joy Johannson, whose mastery of literary English has been an inspiration and a guide. She pushed me to make this a much better work than it might otherwise have been, and greatly increased my pleasure in what I have written. In this as in previous works I have found the advice of my good friend and fellow scientist Robert Eisenstein of immeasurable value.

Contents

Prologue

As promised so clearly and unapologetically in the book of Genesis, knowledge carries with it a terrible burden. Human beings, uniquely among all living creatures on this earth, know that one day they will die. It is a painful knowledge. We have spent most of our history as a knowing species devising belief systems that help us either accept or deny that single fact. No human culture ignores it. It colors our experience as individuals, and often influences our collective actions. Death is a subject that simultaneously terrifies us and fascinates us. Understanding that terror and fascination is an important part of human psychology.

While we continue to think about death from philosophical, cultural or religious points of view, we also study it scientifically. Thanatology, the study of death and dying, is a recognized branch of medicine, with its own scientific journals. But thanatology focuses on the psychosocial aspects of death and dying; it does not ask questions about the nature of death itself. The branch of medicine called pathology describes in great detail the changes in the body and its cells and tissues that lead to or accompany disease and death. A pathologist can tell us whether a tissue is healthy or diseased, alive or dead. But about the precise nature of the razor-thin line separating life from death, the pathologist has little to say.

So *what is death*? One way to understand the death of a human being is to seek the smallest, ultimately indivisible unit — the "atom" of the ancient Greeks — of human life. That unit, that atom of life, is the cell. The cell is the smallest unit in the human body of which we can say, "This is alive!" And if we can define cells as having life, then it follows that it must be possible to describe them in the absence of life — when they are dead. What does a dead cell look like? What is it missing? Why is it dead? How did it make the transition from alive to dead? *How did it die*?

These are important questions, because the death of every human being begins with the death of just a few cells. We normally think of death in terms of *death of the person* — the integrated whole composed of personality, will, memory, passion, and the hundreds of other things that make each of us unique. Most of these characteristics are housed in a specific portion of the brain — the cortex — and the loss of "personhood" that results from loss of cortical function is increasingly viewed as one of the most important aspects of human death. But clearly death must also have a biological meaning independent of the human condition. In the death of our cells, we are no different from all of the other organisms on earth condemned to die as a condition of birth. Snails die, as do worms and mushrooms, and their deaths too begin with the death of just a few cells.

The study of death at the level of individual cells has revealed unexpected subtleties and complexities about the nature of death in multicellular creatures like ourselves — for example, the widespread occurrence of suicide among cells in our bodies. Surprisingly, the study of evolutionarily older single-cell organisms suggests that cell aging and death is *not* an obligatory attribute of life on earth. Obligatory death as a result of *senescence* — natural aging — may not have come into existence for more than a billion years after life first appeared. This form of *programmed death* seems to have arisen at about the same time that cells began experimenting with sex in connection with reproduction. It may have been the ultimate loss of innocence.

Trying to grasp the meaning of an infinite and ever-expanding universe has led toward an enormous abyss in human understanding. From the edge of that abyss, we peer anxiously through our telescopes into the fog of the unknowable. If we travel in the other direction — if we turn inward and, with a succession of ever more powerful microscopes trace the process of death down through the level of individual cells and into the molecules and atoms of which they are composed — we come once more to a fog-filled abyss, one that separates the phenomenon we call life from the cold and indifferent physical universe. And we see through our microscope a figure, peering anxiously at us through a telescope. . . . Death brings us full circle.

Sex and the
Origins of Death

1

Death of a Cell

If you know not how to die, do not trouble yourself. Nature will in a moment fully and sufficiently instruct you. She will do it precisely right for you; do not worry about it.
— *Montaigne*

The average adult human being is composed of something more than a hundred trillion — 10^{14} — individual cells, each with a life of its own. The death of a human being is a direct, irreducible consequence of the death of his or her component cells. But what does death mean at the level of a single cell? And how many of our cells have to be dead before we are dead? In a complex multicellular organism like a human being, are some cells more important for being alive than others? What do we really know about these elusive "atoms of life"?

In fact we know a great deal about the cells that make up our bodies. We know, first of all, that life on earth certainly did not begin in the form of multicellular animals like us. The earth itself came into existence around five billion years ago. The initial atmosphere created by gases escaping from this newly condensed mass was very different from the air we breathe today, and the materials dissolved in the newly formed seas were also very different. The seas contained carbon- and nitrogen-based compounds which could be readily converted, under the influence of the tremendous thermal, electrical, and radioactive energies raging over the earth's early surface, into the basic building blocks of life, such as proteins and nucleic acids. These reactions have actually been reproduced in the laboratory, and the scenarios for explaining how these basic building materials arose are quite believable.

Somewhere around four billion years ago, the very first cells seem to have arisen from this inanimate matter, by processes that today can only be guessed at. The guesses made so far are not very convincing. These early cells did not assemble themselves into multicellular animals for at least two billion years after their first appearance on earth. In the beginning life consisted of nothing more than single, free-living cells. Yet whatever properties we may ascribe to life— the ability to eat, to move about, to produce offspring— were displayed by these single cells. Such organisms still exist today as bacteria, yeast, amoebae, and many other single-cell life forms. Like their forebears billions of years ago, these cells are tough. They have to be. Single cells are extremely small, and now as in the beginning each cell has to survive entirely on its own. Ultraviolet rays from the sun, as well as the oxygen in the atmosphere, pose a constant threat to the very material they are made of. The world around them is dangerous and in a constant state of flux, changing almost hourly. Temperatures shift; food and water sources come and go; the acidity and salt level of their surroundings can wander all over the map, passing in and out of the narrow range able to support life.

The first cells to appear on earth arose directly from materials contained in the "primordial soup" — the collection of bioorganic molecules generated in the high-energy reactions mentioned above. As far as we know, these conditions for producing cells from inanimate matter no longer exist on earth. Cells, whether they are single individuals or part of a multicellular organism, now arise only from other cells. Every human life begins as a single cell, formed by the union of a sperm and an ovum; approximately fifty rounds of cell division are required to produce a fully formed person, by which time the various daughter cells look as different as brain and bone, or heart and bladder. Yet each cell, despite outer appearances, actually differs from every other cell in the body in only the subtlest ways. Each is the end-product of billions of years of evolution, of nature's practice in "getting it right." And each of these near-perfect cells — with one exception — must die. We will discuss this exception a bit later.

The idea that plants and animals are made up of individual cells that correspond to and are ultimately derived (in an evolutionary sense) from the free-living, single-cell microorganisms that still permeate our environment was inspired by the use of increasingly powerful microscopes. The first descriptions of free-living cells like yeast and bacteria, or the amoebae found in freshwater ponds, began to appear in the mid-1600s. They were referred to as "animalcules," or little animals, in recognition of their status as living things. At the time, no one had the slightest idea what cells were, or of their significance as living things or as parts of living things. The notion that the cells that make up plants and animals might also be individual, self-replicating life units took two hundred years to develop, and was not firmly established until the late 1830s, when Theodor Schwann and Matthias Schleiden proposed the "Cell Theory."

With the increasing perfection of the microscope, and especially with the development in the latter half of the nineteenth century of chemical stains that could make the different parts of cells and tissues stand out from one another in sharper contrast, it was gradually realized that the cells making up a tissue have a sophisticated internal architecture, and that this architecture can be related in precise ways to the functions of the cell. The first subcellular structure to be described was the nucleus, which because of its large size had actually been discovered in the 1830s, well before the advent of staining techniques. Identification at the structural level of other parts of the cell took longer, and association of structures with cellular functions only seriously got under way after the development of the electron microscope in the 1930s and 1940s. The major working parts of the cell, called *organelles*, were still being defined into the 1980s. Even today there remain a few structures within the cell about whose functions we are not entirely certain.

Cells are the smallest living units making up our bodies. They are incredibly tiny; ten thousand of them clustered together are just visible to the unaided eye. Yet every cell contains within it, in the form of a molecule called DNA, a kind of chemical hologram of an entire human being. Each cell, at least theoretically, has the information necessary to reproduce the entire being of which it is but the hundred-trillionth part. This has actually been done in the laboratory in a limited way with frogs, and the idea of doing it in humans (and dinosaurs!) has generated more than one science fiction novel. As a practical way of making human beings, however, DNA "transplants" are a long way from becoming a threat to our present means of reproduction.

Most cells in our bodies are born — and will live and die — in complete and utter darkness. The vast majority of cells within our bodies have never seen the light of day. Unless they have managed to get in the way of an X-ray beam, none has ever felt the sting of a photon on its surface. Even the living cells of the skin, buried as they are beneath layers of dead cells, have only a minimal sense of the light, unless we insist on lying under the sun for hours on end without protection. The one exception is cells of the retina that line the back of the eye and gather light from the sun or other stars, or from man-made sources. But this thin cell layer is walled off from the rest of the body by an underlying layer of connective tissue so dense and shiny in some animals that it bounces photons right back through the retina a second time, doubling their rate of capture (a useful trick for night vision). Any photons managing to pass through this connective tissue barrier beneath the retina run into a bone wall — the thick, curving eye socket of the skull. The brain is as much in the dark as any other part of the body.

When life began on earth, cells did not live in darkness, unless they happened to be under a rock or at the bottom of the sea. They certainly did not live buried in a mass of other cells, creating their own darkness. When at last a few cells came together to form multicellular organisms, they unquestionably gained a great deal in terms of security. The inner darkness that comes from being a small part of a large biomass is a great way to escape damage from the sun. Internal environments, particularly in mammals like us, are relatively stable with respect to most of the parameters that sustain life.

But there is a downside to this improved standard of living. Cells that united to become multicellular also became soft. Once they accustomed themselves to their new environment and a life of relative ease, cells lost their toughness, their ability to cope with conditions less than ideal. As a result, human cells are more vulnerable to threats from the environment than are most single-cell organisms. There is a concept in biology and medicine called *homeostasis*, which refers to the delicate physiological balancing act that organisms must perform within the range of temperature, acidity, salinity, oxygen pressure, and other variables necessary for life, and to the ability to control that range in one's environment. The permissible homeostatic range is much narrower in animal cells than in their free-living, single-cell ancestors. Moreover, as fixed parts of a multicellular organism, most have lost their ability to move around if the supply of food or oxygen runs low. They rely on things being brought to them, and on their wastes being carried away for them.

Not only is it dark inside the body; it is also wet. All of our cells are bathed in a gentle, never-ending stream of fluid referred to as *interstitial fluid* or *lymph*. The sources of this stream are the many branches of nearby arteries that bring blood, with its life-giving oxygen and food substances, to every cubic millimeter of the body. Each of these branches keeps subdividing into ever-smaller arteries and arterioles, eventually breaking up into tiny capillaries, microscopic vessels from which oxygen and nutrients diffuse into the surrounding tissue spaces, and from which small amounts of lymph escape to help bathe nearby cells.

In order to understand how cells die, we have to know a little about how they live. We will concern ourselves here with only the broadest outlines of cell structure and function. Imagine for a moment that we are actually inside a living cell — let's say a *myocardial cell,* one of the oblong cells making up the muscular pumping walls of the heart. We will have to bring along some rather powerful lights to see anything inside this cell. We and the lights will also have to be able to work under water — all cells are completely filled with, as well as bathed in, fluid.

Myocardial cells, like many other cells in the body, have their own highly specialized function. Their job is to contract in coordination with one another so as to force blood into the body's circulatory system. Inside each myocardial cell is a set of protein sheets with enormous contractile power. These sheets are anchored like bungee cords to each end of the cell, and occupy over half of its free space. All the myocardial cells making up the heart are under the control of the heart's own built-in pacemaker, the *sinoatrial node.* Sixty or seventy times a minute this pacemaker, assisted by a "booster" — the *atrioventricular node* — sends out a wave of electrical excitation that passes through the individual myocardial cells making up the wall of the heart. For a brief instant, each of these cells contracts its special protein sheets and shortens to a fraction of its normal length. The force of large blocks of cells contracting at the same time causes a contraction of heart muscle, allowing the heart to pump blood throughout the body.

As in most cells, the interior of myocardial cells is dominated by a large, walled-off compartment called the *nucleus.* If you look carefully, you will see what appear to be little round portholes all over its surface. These are where mole-

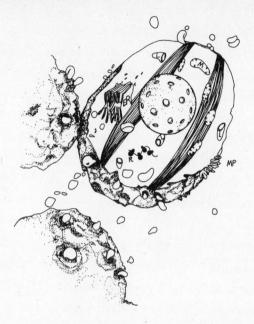

Figure 1. Internal organization of a prototypical contractile cell. In this model cell, a portion of the outer plasma membrane has been removed to reveal the cell's interior. The large centrally disposed *nucleus* houses the cell's DNA. The nucleus is characterized by distinct pores through which molecules pass back and forth between the interior of the nucleus and the cytoplasm. (The pores shown here are over-sized for illustrative purposes.) Bands of contractile filaments stretch from one end of the cell to the other; in a myocardial cell, these filaments would occupy up to seventy-five percent of the cell's interior. Numerous *mitochondria* (M) are disposed around the cytoplasm, many of them in intimate association with the contractile fibers. The *endoplasmic reticulum* (ER) is a major site of protein synthesis in the cell. Proteins are synthesized on ribosomes (R). An enlarged example of ribosomes bound to a strand of messenger RNA (copied from DNA in the nucleus and sent to the cytoplasm) is shown floating free in the cytoplasm. *Lysosomes* (L) are sites for disposal of intracellular waste products. The surface (plasma) membrane of the cell contains numerous *membrane pumps* (MP).

cules are passed back and forth between the nucleus and the rest of the cell. If cells themselves were to have a brain, the nucleus might well be it. The nucleus houses the DNA, which contains (in the form of *genes*) the blueprints for every single characteristic of the cell and the instructions for operating all its machinery. Interestingly, only a few percent of the total DNA in a cell is actually organized into the genes for guiding a cell through life; the rest of the DNA has no discernible function or meaning, and has been labeled "nonsense DNA."

The machinery for operating the cell is found in the liquid *cytoplasm* filling the cell outside the nucleus. These humming oblong tanks over here are actually power generators called *mitochondria*, which convert food and oxygen into the universal currency of energy in living cells, known as *ATP* (*adenosine triphosphate*). If we swing the light over here for a moment, you can perhaps just make out these clusters of slightly asymmetric dumbbell-shaped machines called *ribosomes* strung together by an almost invisible thread of *messenger RNA* (*mRNA*). The mRNA is in effect a xeroxed set of instructions, copied from one of the genes in the DNA, that directs the construction of a protein. The ribosomes operate twenty-four hours a day, seven days a week, stamping out an incredible variety of protein products. Some of the ribosomes float freely in the cytoplasm; others are anchored to convoluted internal membrane structures called the *endoplasmic reticulum*. Most of the proteins produced by the ribosomes will be used by the cell to maintain itself, although some cells—the insulin-producing beta cells of the pancreas, for instance—make proteins for export to other parts of the body. Overhead you can see row upon row of those contractile protein sheets we were talking about earlier, the ones that allow myocardial cells to carry out their special function in heart-muscle contraction. Notice the clusters of mitochondria snugged up next to them for efficient delivery of the large amounts of ATP needed to carry out their repetitive contractions. Be careful—you don't want to fall into one of these structures right over here: they are the *lysosomes*, where all the trash is disposed of. Anything put into a lysosome is rapidly degraded into a soupy mush by powerful chemical agents and potent enzymes.

Finally, as we make our way toward the outer limits of the cell, we will encounter—put your hands out, right over here; you can feel it—the soft, spongy boundary of the cell, the *plasma membrane*. It is made mostly of fat and cholesterol, to keep the watery interior of the cell completely separate from the fluid environment outside the cell. But the plasma membrane is much more than just a barrier. These bumps located every few microns along the wall are actually powerful pumps. Cells depend on these pumps in the same way that reclaimed land near the ocean's edge depends on sea pumps. The environment within a cell is very different from the environment outside. The cytoplasm is jammed full of the special chemicals, proteins and salts the cell needs to sustain life. And the concentration of these molecules inside cells is often much higher than the concentration outside.

Conversely, the concentration of water outside the cell is much higher than the concentration of water inside. As a result, there is a constant tendency for water to rush into cells under osmotic pressure. It is the task of one set of membrane pumps to pump this water back out as soon as it enters. This involves an enormous expense of biological energy, but if it is not done quickly and efficiently the cells will swell and burst. Cells also maintain much lower levels of sodium and calcium ions inside than are found in the surrounding fluids, and much higher levels of potassium ions. Cells use separate sets of energy-driven pumps to maintain these ionic gradients. If any of the pumps shut down, the cell will quickly die. The coordinated activity of these pumps is absolutely vital to the life of the cell.

We can't see them from here, but on the outside of the plasma membrane are all the lifelines the cell uses to stay in touch with other cells. Some of these are simply mailboxes into which other cells deposit chemical messages that are acted on as the cell thinks fit. There are special regions of the cell surface that act essentially as Velcro patches, allowing each cell to adhere tightly to its neighbors. And since we are inside a myocardial cell, we would find just on the other side of this membrane a series of insulated plates through which the electrical impulses generated by the heart's pacemaker reach the cell. Down at the other end of the cell is an identical set of plates where the wave passes through to the next cell. When all is working as it should, sixty to eighty waves pass unbroken through the cell each minute.

Although it doesn't know it yet, the myocardial cell we are in is about to die. It will die because of myocardial *ischemia*, or deprivation of blood supply to the portion of the heart in which our cell lies. The first sign of danger, if our cell could read such signs, is a gradual tapering of the stream of lymph fluid flowing over its outer surface. The ultimate source of this stream — one of the small arterial branches bringing blood to this particular region of the heart — has been gradually narrowing for several years now, like a tiny brook clogged by rocks, tree branches, mud, and other debris. In this case the debris is a complex mixture of fat, cholesterol, and dead blood cells that has been building up inside the arterial wall for several years. This process began when excess dietary fat and cholesterol in the blood were laid down in what is known as a fatty streak, which attracted the curiosity of white blood cells passing through the artery. White cells are constantly on patrol in the bloodstream, looking for anything that might pose a threat to the body. Unable to clear this unwanted material out of the way, they too ended up getting bogged down in the mess, dying and adding to the logjam. As a result the normal healthy flow of blood through the artery has slowed to a tiny stream over the past several months, and the amount of lymph fluid that can be released from downstream capillaries fed by this artery has become vanishingly small.

The cell we are in has had no sense of any of this. But as the supply of lymph bathing the surrounding heart muscle begins to slow to the barest of trickles, and even shuts off intermittently, the cell senses that something is terribly wrong. The decreased flow of lymph fluid means a decrease in the supply of life-sustaining materials dissolved in it, in particular food and oxygen. The mitochondrial ATP generators, responsible for supplying energy to the entire cell, begin to shut down all around for lack of fuel and oxygen. The amount of ATP inside the cell begins to fall below the critical level needed to maintain normal cell function. In response, less efficient backup generators kick in and continue to hum for awhile, burning emergency stores of intracellular food like starch and fat, and even protein, in the struggle to keep up with the demand for energy. But these stores will soon be exhausted, and the auxiliary generators too will be forced to shut down. Momentary metabolic stillness will be added to the dark; in a matter of seconds the lack of ATP will start to wreak havoc everywhere in the cell.

You can probably feel it starting to happen. Most critically affected by the lack of energy are the powerful pumps operating in the plasma membrane at the outer reaches of the cell, the ones that hold potassium in, and keep water and calcium out. So crucial to the life of the cell are these pumps that they are given absolute precedence for the ever-diminishing supply of ATP. It is no longer a question of function; it is now a matter of survival or death. All other energy-driven operations in the cell, including contraction of the sheets that drive the pumping function of the heart, are forced to shut down to save fuel for the pumps. The protein-synthesizing machinery stands everywhere idle in the cell; messages from the nucleus pile up unread. Partially finished products of every description begin to drop off assembly lines as ATP-dependent enzymes wait for new energy supplies to arrive. Chaperones of the unfit and the incomplete rush to transport them to disposal units. The lysosomes are driven to a frenzy as they try to deal with all the trash being fed into them. Everywhere the cry is the same: "Where is the ATP?"

But the ATP never comes; one by one the membrane pumps sputter and lie still. Calcium slips in through the gates that used to exclude it, and begins to corrode and distort the mitochondria bobbing silently in the dark. And then water rushes in, torrents of it. The cell begins to swell, putting unbearable pressure on the outer plasma membrane. Finally this membrane, this wall that isolates and protects the cell from the outside world, begins to crack; the cracks widen with increasing speed until the membrane rips open and the entire cell literally explodes into the outer darkness, spilling its now-useless machinery and its very sap into the nearly dry lymph stream trickling by outside.

These events do not go unnoticed by the rest of the body. The body is a larger community of cells, and like every organized community, it has individuals who specialize in dealing with the dead. White blood cells are constantly on patrol in the body, drifting quietly through the blood and lymph. Some are armed to the teeth, on the lookout for invaders that can cause disease and death. But these warrior cells do not always prevail, and even when they do there may be incredible carnage, with as many dead white cells as dead invaders. So wherever they go, the warrior cells are accompanied by a corps of undertakers called *macrophages*, white cells that may participate in the battle, but who are also trained to take care of the dead. The inner parts of cells floating by in the lymph fluid alert the macrophages to the presence of death, and they begin trudging upstream, working their way through the ever-increasing density of floating debris until they arrive at the source. These dealers in the dead glide silently through the area, probing, testing, sliding on past those with a firm belly, looking for the flabby, waving fragments of membrane that identify the corpses. The blocked artery has resulted in the death of not just one cell, but thousands. There will be much to do.

The macrophages set about quickly and efficiently removing the dead. They do not embalm them, nor do they bury them. They eat them, which is how they came by their name — *macrophage* means literally "great eater" in Greek. They embrace the remaining fragments of dead cells and sweep them inside into their own lysosomes where they are quickly degraded into their component parts, which will eventually be released back into the bloodstream to be used as nutrients by other cells. Thus are the dead recycled inside the body, just as one day the body itself will be recycled in its entirety, through soil and through plants, to provide nutrients and oxygen to nourish human cells yet unborn. The macrophages work silently at their task, recruiting nearby worker cells called *fibroblasts* to help them wall off the area of death with thick layers of pale scar tissue. When all is finished the macrophages will slip away into the lymph to rejoin their warrior brothers, leaving behind a scene bereft of life, as cold and as still and as white as the surface of the moon.

The cell we just watched die is part of a heart, and the heart belongs to a human being—in this case, a man, sixty-two years old. This heart has beat faithfully in his chest over two billion times, sending life-giving blood out to the cells and tissues in his body. But he now lies pale and limp on the hallway floor of his home; he has suffered a major heart attack. It is not his first. His initial heart attack, two years earlier, involved ischemia to a significant portion of heart muscle that subsequently became *infarcted*—converted into dead, functionless myocardium overridden with whitish scar tissue. The pumping efficiency of his heart was reduced considerably, but he was left with enough residual function to lead a fairly normal life. This second attack involves blockage of a different artery, but one that also serves the musculature of the critical left-ventricular heart chamber, which carries the major burden of pumping blood from the heart out into the rest of the body.

He awoke at six o'clock this morning as usual. He sat up in bed and put on his slippers, stood up, yawned and stretched, and headed out of the bedroom to bring in the morning paper. He had just turned the corner into the hallway when he was literally brought to his knees by a horrendous crushing chest pain. There was no doubt at all in his mind what it was; it was like the first attack, but much, much worse. Within seconds he lost consciousness and collapsed the rest of the way to the floor. Like the majority of heart attacks, his came early in the morning, at a time of relative inactivity and low demand on the heart itself.

His wife also knew, within seconds of hearing him cry out and bump against the wall, what had happened. She would recall later that the blood seemed to drain from her completely for a moment, leaving her terrified and helpless. But then, her own heart beating wildly, she took a deep breath, rose quickly from the bed and went out into the hallway. She had tried to prepare herself for this possibility after he had his first heart attack. Warned by their doctor that it might very well happen again, she had taken a course in CPR (cardiopulmonary resuscitation) at a nearby fire station.

Now it is here; now it is real. Pushing panic into the background, she kneels beside him on the floor. He is sweating profusely, his eyes closed. She calls his name, shaking him and slapping his cheeks. He does not respond; he is unconscious. She checks his neck for a pulse and feels none. She knows this is not good, but not necessarily hopeless. She moves quickly to the phone and dials 911. Her voice is shaking and she is incoherent at first. The dispatcher works her calmly through the necessary information. Learning that she is familiar with CPR, he urges her to begin immediately. Help is already on the way.

She struggles to turn her husband over on his back. She cannot detect breathing; there is no rise and fall of his chest, and when she tilts his head back and opens his mouth, she can feel no breath on her cheek. She immediately gives him two of her own lungsful of air in mouth-to-mouth respiration. She moves over him and feels for the tip of his sternum—nearly two minutes have now passed since she heard him fall. She begins a series of rhythmic, rapid downward thrusts with the heel of her palm, three finger widths above the tip of his sternum, to push the blood out of his heart and into the arteries. Alternately with these compressions she forces air into his lungs with her own breath. She keeps repeating this cycle—fifteen compressions, two breaths—until the first-response team arrives four minutes later.

As in most communities, the first-response team is an engine company with firemen trained in basic life-support techniques. Two of the firemen take over administration of CPR, while a third hooks up a heart monitor to the man's chest and a fourth leads his wife into the living room, where he tries to calm and reassure her, and to gather basic information about her husband's health. A rapid assessment of cardiac function indicates that the man is in *ventricular fibrillation*. Electrical signals emanating from the sinoatrial node are coursing throughout the heart in a completely uncoordinated fashion, trying to get the muscles to contract and pump blood. The combination of previously weakened heart muscle and the damage from the current attack has caused his heart to contract spasmodically, in different places at different times, without the integration needed for effective pumping. As a result, there is no regular pulse or recognizable pattern of spiking on the ECG (electrocardiogram) monitor now attached to his chest. The blood flow from his heart to the rest of his body has dropped to a mere fraction of normal.

Virtually all first-response teams are now equipped with a portable electric defibrillator. Experience of the past ten years has shown that for patients in ventricular fibrillation, immediate electrical defibrillation, before administering drugs or any other resuscitative measure, is the most important parameter for saving lives. It has been nearly eight minutes since the beginning of his attack. Paddles are pressed firmly onto saline-soaked gauze pads placed over each side of his chest — one just to the right of his sternum, the other just to the left of his left nipple. A terse command is given and everyone steps back. The man arches in a powerful spasm as 50,000 watts of power surge briefly through his thorax, and then he sinks back to the floor. The purpose of so much electrical power is not to "jump-start" his heart but rather to shut it down completely; when it restarts on its own, there is a good chance the sinoatrial and atrioventricular nodes will be able to reestablish a coordinated cardiac rhythm.

But a glance at the monitor shows the same erratic pattern as before. The defribrillator, its paddles serving as electrodes to the man's body, analyzes the heart rhythm and the resistance of the chest cavity to the electrical shock just given, and adjusts output automatically if further shocks are indicated. They are. After the command to stand clear is given once again, current is quickly applied a second time, and then a third, before something approximating a normal heart rhythm begins to appear. The defibrillator screen flashes a message that no further shocks are needed. One of the firemen keeps an eye on the monitor while another continues to blow through a mouth tube into the man's lungs; he is still not able to breathe normally on his own.

By now the advanced cardiac life support (ACLS) unit has arrived. It is twelve minutes since the attack began. They walk straight through the door purposely left open for them; the fireman in the living room nods toward the hallway. The ACLS paramedics take over smoothly from the firemen. One keeps an eye on the monitor while a second probes and pats the man's veins to find one suitable for an I.V. (intravenous) needle. A third begins to slide a long, slightly curved tube down the man's throat. Manual CPR and defibrillation have failed to restart a normal breathing pattern. It is difficult to get the tube placed properly; although unconscious, the man is gagging and has vomited. Fortunately there is little in his stomach, but it is still hard to get the tube inserted. The paramedic wants to place the tube into the trachea in order to deliver air directly to the lungs. The attempt is interrupted while one of the paramedics uses a balloon pump to force more oxygen into the man's lungs. The intubation requires at least two additional minutes to complete. Finally the tube is properly placed, and the paramedic begins pumping large quantities of pure oxygen rhythmically into and out of the man's lungs. Fifteen minutes have now elapsed.

At last the man picks up the breathing rhythm on his own. His heart pattern appears to be stable. The ACLS paramedics administer drugs through the i.v. line started earlier, to help stabilize him for the ride to the hospital. The breathing tube is left in place; all the other equipment is cleared away. He is lifted onto a gurney, and quickly wheeled to the ambulance. The emergency equipment is thrown into the back, and the ambulance sets off, siren screaming in the still morning air. A neighbor follows in a car with the man's wife. From a telephone inside the speeding ambulance the paramedic in charge describes the situation to the hospital so that staff will be ready when they arrive.

Many questions remain, and they can only be answered by coronary critical-care specialists who will perform the necessary evaluative and laboratory tests at the hospital. Unquestionably, had he not received immediate CPR followed by defibrillation and intubation, their patient would have been dead many minutes ago. But he is still unconscious, which worries the paramedics. And he was not breathing properly when they arrived at the house. Did he manage in spite of his trauma and breathing difficulties to get enough oxygen to his brain cells to prevent irreversible brain damage? How much cardiac damage — infarction of heart muscle — has this morning's attack inflicted? Will his heart be able to withstand much longer the cumulative destruction of two major attacks?

We will rejoin our patient a little later, once he has reached the hospital. He will be subjected to further emergency procedures to stabilize his condition, and then examined by specialists to determine the precise extent of the damage. In the meantime, let us examine death a bit more closely, for it is a possible outcome of our story.

2

A Second Face of Death

*Let us therefor consider how farre and in how many waies
selfe-homicide may bee allowable.*

—John Donne

The death of a myocardial cell described in the last
chapter—caused by starvation, and by suffocation from lack
of oxygen—was an ugly death: violent, chaotic, disorderly.
That kind of death happens as a result of accident or mischance, and is called *necrosis*. It is the way death comes to
fragile cells inside the body when the delicately balanced
conditions for maintaining life veer too far off center. It is
also the way cells die when they are poisoned, for example by

one of the many toxins released by bacteria and other pathogenic microorganisms. Necrotic cell death is usually accompanied by wholesale destruction of the internal parts of the cell, by the gushing in of extracellular water and the ripping open of the cell membrane. The macrophages that come in to clean up the dead and dying cells trigger the deposition of scar tissue that makes it difficult for healthy cells to move into the damaged area and replace lost cellular functions. Scar tissue is a useful response to damage in many parts of the body; it promotes healing and keeps tissue from losing its shape. But it cannot help a heart beat or a brain think.

For many years, this form of cell death was thought to be the way all cells came to an end. In fact, no one paid much attention to how cells died. Perhaps understandably, biologists mostly have been fascinated with how cells live, how they function, how they reproduce themselves. But cell death can be equally complex and fascinating. It turns out that there is another way entirely in which cells die, a way that is very different from necrosis. It is not a result of accident or mischance; it is programmed into cells and is activated only under very special conditions. Its discovery led to the creation of a whole new field of inquiry, and now the study of *programmed cell death* is one of the leading topics of molecular biology and medicine.

To examine this second face of cell death, let us enter for awhile the world of a human fetus developing in the womb. This would seem an unlikely place to seek death, this very temple of life and growth and increase. Yet here, too, death plays a role—an important role, one absolutely vital to the creation of a complex new being. One of the instances in which this can be seen most readily is in the genesis of the human hand.

In the first eight weeks of life, human embryos undergo almost uninterrupted cell growth. This is the period during which the plan for the entire body, including all the major internal-organ systems, is laid down. By the end of the eighth week a human embryo is clearly recognizable as human, and it graduates to the status of fetus. The limbs of a human fetus—the future arms and legs with their appended hands and feet—make their first appearance during the period of embryonic growth, at the end of the fourth week of life. They begin as small bumps on the margin of the evolving body structure, pushing out rapidly over the next several weeks to achieve their final form. The future arms are always a few steps ahead of the future legs. By the end of the sixth week of development, the three major segments of the arm are clearly visible: the upper arm, the forearm, and the hand.

The hand at this stage looks more like a ping-pong paddle than a tool that will one day hold a pen or a violin bow. Traces of future finger bones are just barely discernible as faint lines of condensing cartilage connected by webs of tissue. This is a characteristic stage in the development of all vertebrate animals, and is an example of the embryological principle laid down by the biologist Ernst Haeckel in the last century: *ontogeny recapitulates phylogeny* (the history of an in-

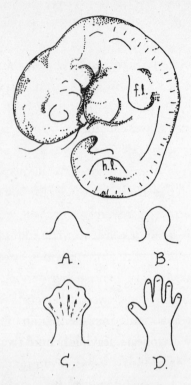

Figure 2. Development of the human hand. The human forelimb first appears as a tiny swelling along the trunk of the fetus at the end of the fourth week of development. By the fifth week the forelimb (fl) has begun to push out (top figure and a), and the hindlimb (hl) bud has appeared (top figure). At the end of the fifth week the forelimb is a simple paddlelike structure (b). At six and a half weeks (b), the cartilage that will give rise to the bones of the hand and fingers begins to condense, but the future fingers are still connected by a webbing of cells (c). By two months (d), the hand is fully formed; the fingers even have tiny nails.

dividual fetus in the womb recapitulates the biological history of that fetus' ancestors). Although stated somewhat overambitiously by Haeckel, the principle has a certain validity. All vertebrate embryos, for example, pass through a stage where they have gill structures in the neck region. Fish keep these gill structures to help them breathe underwater as adults. Human and other higher vertebrate embryos pass on through this stage, using leftover gill tissue to fashion more useful structures, like the thymus or the thyroid glands. Similarly, all vertebrate embryos pass through a stage where the digits of the hands and feet are webbed. Fish and some birds maintain this webbing throughout life, reinforcing it and using it in the construction of fins or wings or webbed feet. In human embryos, between the forty-sixth and fifty-second days in the womb, the interdigital webbing of the hand suddenly disappears, leaving behind five beautifully shaped fingers. Trailing along just a few days behind, the same process is repeated to create a human foot complete with toes.

It is strange that this process had been described in great anatomical detail for a hundred years or so before anyone bothered to ask what happens to the cells that make up the interdigital webbing. It turns out that they do not just move away to some other part of the body, or get incorporated into the palm of the hand or the nearby wrist. They die. One by one, over the course of a few days, each of the cells forming the webbing between the fingers and toes of a human embryo dies. But they do not die the raucous, violent death of necrosis. They do not die because their blood supply is interrupted or because water or deadly calcium seeps into them. They act in response to a script embedded deep within them long ago, over which they have no control. Responding on cue to signals from their environment, they commit suicide.

The death of a cell by suicide is altogether different from necrotic cell death. Necrotic cell death is *cytocide*, the killing of a cell that does not want, and is not scheduled, to die. The cell dies in response to changes outside itself, some lethal alteration in its immediate surroundings. As we have seen, a cell dying of necrosis struggles violently, with everything at its disposal, to avoid death. The act of suicide by a cell is completely different. One cannot help being struck by how peaceful a death it seems, and all in all a rather unmessy one. Although it may involve large numbers of cells at a given location, the cells do not explode as a result of osmotic imbalance. There is no rushing in of water, or spillage of intracellular debris into the surrounding tissue fluids, attracting the eaters of the dead.

It would be easy to anthropomorphize the suicide we see in cells, and we mustn't do that. Suicide by cells lacks completely one very important element of suicide as we commonly think of it: volition. Individual cells do not have anything remotely approaching free will either in their lives or in their deaths. When cells commit "suicide," they are responding to a program they cannot alter in any way. The impulse to commit suicide ultimately does come from within, but it is not in response to sorrow or despair, nor is it a form of altruism. None of these exist in cells. What does exist, in all cells — in every cell in our bodies — is a built-in program for self-destruction, should the need arise. The number of situations in which the need arises is surprisingly large.

One of the first events to occur in most cell suicides is a subtle one, yet for a living cell it is as final and irreversible as any act of self-destruction could be. As we have seen, the nucleus of a cell is in a sense the nerve center of the cell. It contains in its DNA the master plans for every protein the cell is equipped to make, along with complex sets of instructions for regulating their production at just the right time. Virtually every aspect of a cell's life is regulated by its DNA, including its death. Once a cell commits itself to death by suicide, it copies off one last set of instructions from the DNA in the nucleus and sends them to the machinery located out in the cytoplasm. These are the instructions for the cell's own death. As soon as they have been received and processed, the cell begins to destroy all of the DNA in its nucleus. The DNA is broken up into millions of tiny bits that can no longer convey any useful instructions to the cell, in the same way that this page, if torn into a thousand minute fragments, could no longer be read. This does not happen because some deadly destroyer of DNA has managed to penetrate the nucleus; everything necessary to do the job is already in the nucleus, waiting for just the right time, waiting for a signal to set the process in motion. From the moment its DNA has been destroyed, the cell cannot reverse course; it cannot alter its decision. It may take awhile to become apparent, but the cell is already dead.

The fact that its central command system has been destroyed is not immediately noticed by the rest of the cell. There is always a stack of instructions from the nucleus piled up in the cytoplasm, waiting to be read, and the cell can continue to operate for quite some time after destruction of its brain — of its DNA. It carries on for awhile clearing away the backlog of work to be done, including processing of the death messages. Its situation is somewhat analogous to that of a human being who is brain-dead, but whose body, with

Figure 3. Death by apoptosis. The cell shown here was originally a simple sphere, surrounded by a well-defined plasma membrane. It has now begun to disintegrate into numerous smaller apoptotic bodies, each of which contains a small portion of the interior of the previously intact cell. Redrawn from an electron micrograph of a real cell undergoing apoptotic death.

modest assistance, can continue to function more or less normally for weeks or even months. But as surely as a patient who is brain-dead will never again be able to participate in those functions we associate with human life, a cell whose DNA has been destroyed is irreversibly, irrevocably dead.

If we were to view this process from outside the cell, we would have no indication that the cell's DNA is gone, or that anything is amiss. The first hint that something unusual is underway involves the cell's plasma membrane. As a sign that a given cell has somehow been singled out for a fate different from its neighbors', the cell physically detaches itself from them. One by one it breaks the points of contact between its own plasma membrane and the membranes of surrounding cells, until it stands alone. And then the cell begins a slow dance of death; its membrane begins undulating to and fro, ruffling like the gossamer tissues of a Portugese man-of-war propelling itself through water. Portions of the plasma membrane surge out and then fold back on themselves. Small pieces of the cell begin to pinch away from the main cell body, and float idly in the currents of the surrounding lymph.

The anatomical events accompanying suicide in cells were first described in detail by three Scottish scientists at the University of Aberdeen in 1972. In consultation with a classics scholar at their university, they came up with a wonderfully appropriate name for it: *apoptosis* (ap'o-to'sis), a Greek word that describes the falling away of petals from a flower, or leaves from a tree. That is exactly what cell suicide looks like when seen through a powerful electron microscope. Portions of the cell, with various bits of the internal cell machinery — ribosomes, mitochondria, even lysosomes, still surrounded by plasma membrane — fall gently away from the main body of the cell. Inside these cell fragments, or *apoptotic bodies*, as they are called, life seems to go on more or less normally. Ribosomes, if they are included, continue to make protein; entrapped mitochondria still churn out ATP; pumps at the edges of the membrane still labor to push excess water out of the cell. It is as if, at least for awhile, all the systems trapped inside the apoptotic bodies are completely unaware that they have been partitioned off from the main body of the cell.

The calm prevailing inside the apoptotic bodies is reflected in the surrounding tissue spaces. There are no outward signs of death—no ruptured membranes, no chemicals and debris from exploded cells washing downstream in the lymph. There are no legions of white blood cells marching into the area, looking for the killing fields. The apoptotic bodies are eaten quietly and efficiently by neighboring cells, not by professional undertakers. Of course, if a macrophage happens by, it will join in and help speed the process along. But the macrophage does not send out alarms to attract more of its kind to the scene. Nor does it — and this is very important — stimulate nearby fibroblasts to lay down scar tissue. Areas of the body in which cell suicides have taken place are not walled off or given up as dead and useless. Cells disappearing by suicide leave behind only normal, healthy tissues, bathed as always in nutrient- and oxygen-filled lymph fluids.

Apoptotic bodies can actually be seen inside neighboring cells shortly after they have been eaten — still intact, still blithely carrying on their business unaware that anything is wrong. Only at the final moment, when they are shepherded into their new host's lysosomes for destruction, do they seem to realize something is seriously wrong. But a brief instant later it is all over, and they are started down the pathway of decomposition and disassembly that will return them to their elements — which of course will be used by the accommodating host cell in pursuit of its own ends.

The death of cells by suicide is involved in a great deal more than just the shaping of fingers from a webbed hand. In the developing human fetus, cell suicide also plays a major role in the formation of the nervous system. Nerve cells in the brain and spinal cord (*neurons*) are connected to other parts of the body by nerve fibers — long, thin extensions of cells residing in the brain or spinal cord and carrying electrical impulses that stimulate targeted cells to perform specific functions. At a certain stage of fetal development, these neurons begin generating enormous numbers of nerve fibers, which they simply cast out in the general direction of tissues and cells needing nerve connections. If a particular nerve fiber happens to find a cell with a nerve attachment point on its plasma membrane (a muscle cell, for example), it makes a connection. That fiber (and the brain or spinal-cord neuron from which it came) survives and becomes the nervous system's communication line to the targeted cell for life. If, on the other hand, the nerve fiber fails to establish contact with an appropriate cell — and fewer than half do — the neuron that sent it out must commit suicide, dying the same quiet apoptotic death that helped to form the hand.

The role played by cell suicide in the genesis of the nervous system represents an interesting and fundamental fact about the biology of this kind of dying in many cells: death is actually the default state for each of these neurons. From the moment a neuron is spun out of the central nervous system toward potential target cells, it is destined to die. Only if it finds a connection with another cell will it be rescued from an otherwise certain death; it will receive chemical substances (called *growth factors*) from the target cell that in effect switch off the death program. In some respects this seems an incredibly wasteful way to build a nervous system. Each nerve cell that fails to make a connection with another cell in the body, and thus goes on to commit suicide, was very expensive to make in terms of biological energy. As with other tissue-shaping processes in which apoptosis plays a role, the development of the nervous system probably reflects a phylogenetically earlier process that was more efficient. Although now it may be considerably less efficient, overall it must be less expensive and more practical to use the inherited system at a lower efficiency than it would be to design a completely new way of building a nervous system from scratch. So the brain and the spinal cord build millions of cells they will never use, shunting the unselected into death through suicide.

The process of suicide continues after birth and throughout life. Cells of the immune system, including the warrior cells we met earlier, are also generated in great excess. These cells — white blood cells called *lymphocytes* — are allowed to circulate throughout the body for several weeks after they are formed. If they encounter a threat to the body—a foreign protein in the bloodstream, or a virally infected cell — and eliminate it during that time, they will be granted longevity. As when neurons make contact with a targeted cell, lymphocytes detecting a foreign protein or engaging a virally infected cell are rewarded with growth factors that switch off their death program. The resulting cell may survive five years, or ten, or even for the life span of its host, providing a type of "memory" of pathogens previously encountered by the immune system. But if they fail to find and eliminate a foreign invader during the allotted trial period, they are invited, in effect, to fall on their swords. Again, death is the default state for these white blood cells. They undergo exactly the same kind of death as cells in the webbing that binds embryonic fingers together. The phenomena of cellular overproduction, selection, suicide, and memory are examples of the many ways in which the nervous system and the immune system seem to parallel one another.

There is another interesting situation in which cells of the immune system commit suicide: if they receive too much radiation. This phenomenon has long intrigued and puzzled both immunologists and radiation biologists. One characteristic feature of the immune system is the tremendous amount of cell division that is constantly going on in order to satisfy the body's demands for fresh supplies of immune cells. Part of this demand is of course created by the fact that so many immune system cells end up committing suicide when they fail to find anything foreign or abnormal to attack. Cells that are constantly dividing are particularly susceptible to radiation damage; that is the basis of radiation treatments for cancer. Radiation is dangerous because it can introduce mutations into the DNA of cells that are dividing. Mutations creeping into cells of the immune system can be particularly dangerous, because they may cause the immune system to attack normal, healthy cells as well as cells compromised by pathogens or cancer. This possibility preys on one of the body's most deeply seated fears: the development of abnormal DNA — DNA that is not self; DNA that could turn against self. To avoid this risk, it seems that the immune system chooses to rid itself of radiation-damaged cells, and again, the preferred death in such situations is suicide. Interestingly, most current treatments for cancer play on this same fear. Both radiation therapy and chemotherapy act by damaging a cell's DNA, ultimately causing tumor cells to undergo apoptosis.

In discharging its duty to rid the body of intracellular predators, the immune system also exploits the propensity of other cells to commit suicide when their integrity is compromised. This is the specific task of a highly specialized cell called a "killer T cell" around the lab. (The more polite term *cytotoxic T lymphocyte* or *CTL* is used in scientific papers and grant applications.) CTLs are one type of the warrior white blood cells we met earlier, the ones that patrol the body with undertaker macrophages in tow. One of two major arms of the immune system, they were discovered in 1960 in connection with organ transplant rejection. To this day they remain the primary immunological barrier to this potentially life-saving procedure.

It was recognized early on that rejection of organ transplants could not be the CTL's *raison d'être*, but it was not until the 1970s that the real task of the killer T cell was discovered: guarding the body against cancer, and ridding the body of viruses. Some pathogens, like bacteria, are found swimming freely in the blood or lymph; these are dealt with by the other major arm of the immune system: *antibodies*. Antibodies are specialized proteins that circulate throughout the body and bind to pathogens, leading to their rapid elimination. But in some cases, viruses and other pathogens may actually invade a living cell inside the body, taking over the cell in order to reproduce themselves. This can pose a serious threat to the rest of the body, particularly if the pathogen involved is able to spread from one cell to another. Left unchecked, such pathogens can lead to the functional or even actual loss of large blocks of vital tissue.

Because these pathogens hide inside the cell, they cannot be "seen" by antibodies patrolling in the lymph fluids outside the cell. Killer T cells are not so easily fooled. They patrol the entire body, using the same highways and byways of the bloodstream and lymph used by antibodies. But CTLs are equipped with special sensors that allow them to detect the presence of intracellular pathogens. They can spot such pathogens because all cells in the body display on their surfaces samples of the proteins produced inside. If a foreign invader is hiding inside a cell, using the cell's machinery to make its own proteins, samples of those proteins will eventually find their way out onto the cell's surface. CTLs know exactly which proteins are normally made by the body and which aren't. If a CTL encounters a cell displaying a foreign protein on its surface, it will instantly kill it — no questions asked; no quarter given. Antibodies cannot do this. (The cells within a transplanted organ are covered with foreign proteins, which is why transplants are so vigorously rejected. It doesn't matter that the transplant could save the recipient's life; from the CTL's point of view, it is not self — it must be destroyed.)

But how do CTLs do it? How does a killer T cell, once it decides a given cell in the body is somehow compromised or foreign, kill that cell? The assumption from the beginning had been that the CTL must be doing something proactive to cause the selected target to die. Scientists were looking for a smoking gun or a bloody knife, for a rope or traces of a poison — anything that might have been used in the lethal hit. But no matter how hard they looked, or how long they studied the process from start to end, no truly believable weapon could be found.

And then one day a dozen or so years ago, someone decided to take a closer look at the target cell in its death throes, just after it had received the "kiss of death" from a killer T cell. CTLs were mixed with target cells in culture dishes outside the body, and followed by microcinematography. Enlarged images were projected on a screen, run forward and backward; sped up and slowed down. As expected, the CTLs approached the targets, bumping and probing and then locking on tightly for several minutes. But as the target cell was released from the CTL's embrace, it began to do what everyone suddenly realized was a classic cellular dance of death. Its membrane began to ruffle and blister, releasing small particles that looked very much like apoptotic bodies. And these bodies did not explode, spewing their contents into the culture dish; they just sat there, bobbing gently in the nutrient medium. This sent scientists racing back to the lab to monitor the state of the target cell's DNA during the killing reaction. They were astonished to find that within seconds of the lethal hit the DNA was gone—fragmented into a million tiny, useless pieces. And the target cell itself fell gently apart into apoptotic bodies. There was no doubt about it. Cells selected by CTLs for death are not murdered; they commit suicide.

So it turned out that all of the years spent looking for special CTL weapons had, after all was said and done, been wasted. CTLs are not equipped with weapons for destroying altered cells. What they are equipped with is knowledge of a special security code. *Every cell in the body*—not just a few

43

extraneous cells in the developing fetus—has embedded in it a self-destruct program. What CTLs know, uniquely among all the cells in the body, is how to punch in the security code that activates that program and ultimately causes the selected cell to commit suicide. The Kevorkian option, you might say—a kind of assisted suicide. It is absolutely vital to the overall health of a multicellular organism that any cell compromised by radiation or by pathogenic infection be eliminated as quickly as possible, with as little damage to surrounding cells as possible. Of the two types of death we have seen so far, suicide is far the preferable option from this point of view.

There is, however, one slight drawback. CTLs are very effective in seducing compromised cells to commit suicide, but they are essentially blind in their seduction. A CTL is not equipped to make judgments about whether the existence of a strange protein on the surface of a cell truly signals a life-threatening situation for the organism as a whole. In the majority of cases non-self proteins on a cell's surface spell trouble, and the resulting CTL-induced suicide and loss of cells is amply justified. Organ transplantion is clearly one exception, although we can hardly fault the immune system for trying to do its job in a biologically abnormal situation created by humans. But there are other exceptions, not at all man-made, and some of them can lead to disaster. For example, there are a few microbes that can invade cells and live there without causing disease; that is to say, they are nonpathogenic. If a few cells were lost in the course of eliminating such harmless microorganisms from the body, the damage done would be of little consequence. But on occasion, for one reason or another, the CTLs are never quite able to eradicate these harmless microbes completely; they may be particularly well-concealed, or they may spread from cell to cell more rapidly than CTLs can chase them down. The problem is, the CTLs never stop trying. Every time they encounter a cell displaying proteins from the innocuous microbe, the cell is ordered to commit suicide.

In some cases this can take a terrible toll on the body. Consider, for example, hepatitis B virus (HBV) infection in humans. HBV-induced viral hepatitis (also known as serum hepatitis) is roughly the modern equivalent of the black plague. It affects more than 300 million people worldwide, and is one of the world's leading causes of death from infectious disease. HBV induces both an acute and a chronic form of hepatitis, and is a leading cause of liver cancer. The initial symptoms of HBV infection are usually unremarkable, barely more than a mild influenza. When HBV invades cells, it integrates its own small piece of DNA into the infected cell's DNA. Once this happens, the cell treats the viral DNA just like its own. It copies out the HBV instructions for making proteins used in the construction of more HBV, and sends these instructions out to the cytoplasm for processing. In the course of this activity some of the viral proteins make their way out to the cell surface.

The response by CTLs is immediate and vigorous. The infected cells are quickly induced to commit suicide; in the acute form of the disease the infection is often completely cleared on this basis in fairly short order. The resultant damage to the liver may be severe, but it is reparable and only rarely fatal. But in a certain number of cases the disease is not resolved at the acute stage, and progresses into a more chronic form of HBV hepatitis. This is where the greatest damage is done. The viral DNA continues to direct production of viral proteins, which continue to make their way out to the surface of infected liver cells. And the CTLs continue to induce the infected liver cells to commit suicide. Because the liver has a certain capacity for self-regeneration, it keeps trying to replace the missing cells with new ones. But these too become infected as HBV spreads slowly throughout the liver in an ongoing cycle of death and renewal. Over time the renewed, virally reinfected liver tissues become more and more abnormal, failing to carry out even routine functions such as metabolising food and manufacturing blood coagulation products and bile. In some cases, cell loss simply outpaces renewal, leading to liver failure and death for the patient. In other cases the constantly replicating liver cells become cancerous and start to grow rapidly and without control. In a high percentage of advanced cases, particularly in Third World countries where the necessary intensive care is unavailable or inadequate, the result is death. And yet, as far as we know, the hepatitis virus itself is absolutely harmless; if put into culture with human liver cells it will infect them, but the cells remain perfectly healthy.

There are other examples of CTL-induced mass suicides

that result in serious disease or even death for the host organism. In most cases the pathogen, if left alone, would probably kill the host anyway; the suicides prompted by CTLs simply hasten the process. But in other cases, like HBV viral hepatitis, the pathogen itself is harmless; the damage is done entirely by the immune system. This sort of *immunopathology* is turning out to be much more common than was originally thought, and may well underlie a wide range of human maladies previously considered of uncertain origin.

So the phenomenon of cell suicide, which plays a peaceful and positive role in the shaping of embryonic tissues, is also used to harsh but necessary advantage by killer T cells in adults. Killer T cells themselves, if they fail to find an infected cell to goad into self-destruction, must also commit suicide to make room for new T cells with better instincts. These are all deaths programmed into the overall life plan of individual animals; they are carried out to give the individual a chance to stay alive longer and to produce more offspring. The suicide of cells turns out to be almost commonplace, a perfectly normal — and essential — part of the rhythms of animal life.

The origins of cell suicide lie very deep in our evolutionary past. Consider one of the tiniest of multicellular animals: the primitive worm *Caenorhabditis elegans. C. elegans* is so small that every single one of its cells can be counted and accounted for, during its entire lifetime. Like all living things, it starts as a single cell — an egg — and divides until it becomes an adult. But *C. elegans* as an adult comprises only 959 cells. Not 955 cells, not 962 — 959, exactly. Of these, precisely 302 cells form its tiny nervous system. That number of cells in a human being would not even show up as a wrinkle on an eyelid.

For *C. elegans*, it is only twenty-one days from conception to death of the organism. But during the transition from egg to adult, an additional 131 cells are generated that never make it to the adult stage. Over a period of precisely seven hours, these 131 cells commit suicide. Why they have to die is not at all apparent. Were they part of the pattern for some structure even more ancient than *C. elegans,* a structure no longer of use to the adult worm? We do not know what these cells might have become had they survived. But how they die is perfectly clear. The sequence of steps followed by these cells is identical to those taken by cells marked for self-destruction in human beings: the destruction of DNA in the cell nuclei, the detachment from their neighbors, the brief dance of death, the formation of apoptotic bodies that are engulfed by neighboring cells. The ritual of cell suicide appears to be extremely ancient, and almost perfectly preserved across eons of biological time.

The sameness of apoptosis across such a wide range of biological situations and evolutionary time, and in the purposes for which it is used and the steps involved in its execution, led scientists to formulate the notion of *programmed cell death*, or PCD. The events that trigger PCD are almost always the same. Anything that damages a cell's DNA beyond the point where it can be readily repaired will quickly induce it to commit suicide. Cells also commit suicide in response to a wide range of extracellular stimuli. CTLs are one obvious example of such external signals, but apoptosis induced by CTLs represents a fairly recent co-option of PCD by the immune systems found only in humans and other vertebrates. The evolutionarily older use of PCD (also shared by vertebrates, of course) is largely in response to extracellular chemical signals coming into the cell through the plasma membrane. As we saw previously, among the structures embedded at the cell surface are "mailboxes" into which a wide range of chemical messages can be deposited. These messages, collectively referred to as *cytokines*, come from other cells in the body — sometimes from the brain, sometimes from hormone-producing endocrine glands, and sometimes just from a neighboring cell. This "chemical chatter" is how cells in large multicellular animals stay in touch with one another. But cytokines do not always just represent idle chatter; they can be the difference between life and death. Many cells require constant receipt of vital cytokines, such as the growth factors mentioned earlier, to stay alive. If the growth factors never arrive, or are withdrawn, the cell will undergo apoptosis. Cells that do not require constant supplies of growth factors to survive may nevertheless be induced to commit suicide by deposit of a specific instructional cytokine into a plasma membrane mailbox.

All of this is carefully regulated in the body, ultimately in keeping with programs embedded in an organism's DNA; hence the term "programmed cell death." Scientists are now beginning to identify some of the genes that control these programs. *C. elegans* has been a particularly powerful tool for unraveling the genetic basis of PCD. There are genes in *C. elegans* that specifically encode death messages, and cell-surface cytokine receptors (mailboxes) that recognize only those messages. For cells with death as a preset, time-dependent default state via PCD, the timely arrival of a critical growth factor can often override an otherwise certain extinction. There are also genes in *C. elegans* that encode messages capable of reversing the process of PCD once it has been set in motion. These rescue messages don't arrive in a mailbox; they are produced within the cell itself, apparently to give the cell an opportunity to save itself, for example in the face of relatively minor damage to its DNA that may have triggered apoptosis. If the DNA can be repaired quickly, the cell may be allowed to live. But in the face of continued indications that the cell has been compromised — massive DNA damage, the prolonged absence of a required growth factor, or constant bombardment with a message to commit suicide — the rescue message is drowned out, and the cell starts down the irreversible pathway of apoptosis.

These same messages, and the genes that encode them, are now being identified in humans, and perhaps not surprisingly they are only modestly different from those found in *C. elegans*. The battery of genes and chemicals involved in the generation, delivery, and processing of death signals in animal cells is impressively large; programmed cell death is obviously of great importance to nature. We shall return to a discussion of this apparently universal form of cell death, and how it relates to our own mortality, after we ponder an even larger question: where did obligatory, programmed death come from?

3

Sex, Segregation, and the Origins of Cellular Death

For man that is born of woman is of few days, and full of trouble. He flees like a shadow, and continues not.
                                        —*The Book of Job*

Why death?

From the descriptions in the preceeding two chapters of some of the ways in which cells die, we begin to get a sense of *what* death is at the cellular level. Like the death of the being of which it is a part, the death of a cell is a return to unconnectedness, to chaos, and to silence. But *why* do cells die in the first place? Is there something inherent in the nature of life that requires all living things to die? To get a sense of the answer to this question, we must go back in time to the first appearance of cells on earth.

The very first life forms, as we have seen, were not animals in the ordinary sense of the word, but free-living individual cells we now call bacteria. The entire being in this case consists of just a single cell. Yet by any biological criterion that would define us as alive, so were they. The earliest of these organisms represented then, as now, the simplest possible structure for carrying out the cardinal function of all living things: the reproduction of their own kind through replication of their DNA, and transmission of that DNA to offspring.

But it is less obvious that the earliest forms of these single-cell organisms shared then, or share now, the second cardinal feature of life as we know it — obligatory, programmed death. We, like virtually all other multicellular animals, *must* die, and there are many mechanisms built into us to be sure that we do. Some insects measure their life in days; mice live three years if they're lucky. Humans on rare occasions may survive to 120 years, some turtles to 200. But all animals eventually die. Many single-cell organisms *may* die, as the result of accident or starvation; in fact the vast majority do. But there is nothing programmed into them that says they *must* die. Death did not appear simultaneously with life. This is one of the most important and profound statements in all of biology. At the very least, it deserves repetition: *Death is not inextricably intertwined with the definition of life.* Where, then, did death that cannot be avoided come from? Almost certainly, it did not arrive on the scene until a billion or so years after life first appeared. To understand how programmed death came about, it will be useful first to take a brief look at how the earliest single-cell organisms evolved into more complex life forms.

Of the five kingdoms into which all living things are commonly divided, single cells comprise one kingdom in its entirety, and are a significant proportion of two others. The earliest single-cell organisms, from an evolutionary point of view, are the bacteria, and these make up the first kingdom, the kingdom of the *Monera* (See accompanying Figure). Monerans have been enormously successful; they are still around after several billion years of competing with each other, and with more complex life forms. Today they account for approximately half of all the biomass on earth. They inhabit planetary niches extending from frozen arctic tundra to boiling sulfur vents at the bottom of the sea. The bacteria inhabiting the world today provide a window to the past that allows us to guess what the earliest life forms must have looked like, although clearly the ones we study now must be considerably different from the first forms that appeared nearly four billion years ago.

Bacteria came into existence when the earth's atmosphere was largely devoid of gaseous oxygen. The gradual increase in the concentration of oxygen in the atmosphere that occurred around two billion years ago (caused by hydrogen-hungry, water-splitting photosynthetic *cyanobacteria*, some of which would later become parts of plants) was an evolu-

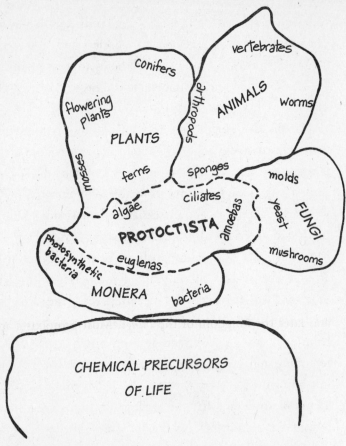

Figure 4. Major evolutionary groups. All classifications of living things are necessarily artificial, and subject to constant debate and revision. The scheme shown here is one of the simpler evolutionary "trees" for the major organismal groups. Protists are the smaller protoctists.

tionary event of major proportions. Oxygen is a deadly, corrosive gas, causing iron to crumble into rust and wreaking havoc with nearly all of the organic molecules on which life is based. The monerans began to develop specializations to protect themselves from oxygen. They had to. Those that didn't simply did not survive.

Although the vast majority of bacteria live their entire lives as single cells, a number of bacterial species have experimented with multicellularity, which would become the major evolutionary direction for most succeeding life forms. Many bacteria live in simple clusters or colonies, trapping and sharing food. A few, like the *Myxobacteria*, form rather elaborate structures called *fruiting bodies* that look like mushrooms or tiny trees. In general, however, bacteria failed to explore the tremendous advantages of multicellularity.

Some of the more successful monerans went on to form new kingdoms of living things: first the kingdom of the *Protoctista* (the smaller members of which are called *Protista*), and later the kingdom of the *Fungi*. Most protoctists are single-cell organisms, like the bacteria. Their kingdom includes things like the slime molds and amoebae, as well as cells like *Plasmodium*, which causes malaria, and advanced forms of algae, some of which continued the transition from bacteria to plants.

Several important features came to distinguish the protists from the earlier monerans. The genetic material of most bacteria consists of a single, usually circular piece of gene-bearing DNA. It is attached to the inside of the cell's limiting membrane, but otherwise floats free and naked in the cytoplasm; there is no nucleus. The DNA is naked in the sense that it is not complexed with the specialized proteins known as *histones*. Thus, these earliest cells are called *prokaryotes* ("pre-nucleus cells"). Protoctists and all later cells divide their generally larger allocation of DNA into multiple *chromosomes* complexed with histones and stored in a nucleus. (The name of these DNA-protein complexes comes from the fact that they can be stained with certain chemical dyes and actually seen in microscopes as "colored bodies" — chromosomes.) In almost all cases these chromosomes are linear, rather than circular; their tips are capped with special DNA structures called *telomeres*, to keep the ends from sticking to themselves or to other chromosomes. The protoctists and all other nucleated cells are known as *eukaryotes* ("true-nucleus cells"). Eu-karyotes also developed something called *diploidy*: that is, they began to carry two copies of each chromosome, rather than a single copy like the *monoploid* prokaryotes. (One member of each chromosome pair was inherited from each of the cell's parents.) Diploidy has the major advantage that damage to or mutation of a particular gene generally has less severe consequences than in prokaryotes because there are two copies of every gene. It also provides an opportunity for purposeful experimentation by nature with the DNA composition of genes, leading to possible improvement in gene function. If an experiment on one copy of a gene fails, there is always a backup gene to see one through the day.

While remaining single-celled, some of the protoctists grew to be very large. Size has definite advantages. It certainly discourages predators, but perhaps of more importance in the early days of life on earth was the ability to store food inside the cell for use when nutrients in the environment dwindled or disappeared. Protoctists like *Paramecium* are easily a million times larger in volume than most bacteria, providing plenty of room for pantries and larders. And as cells increased in size, they also began to develop a number of architectural specializations designed to help them deal with an increasingly complex world. One early specialization was the development of a true *cytoskeleton*, intracellular protein rods that not only helped the larger protist cells maintain their shape but also came to play a role in locomotion and in feeding. For instance, protoctists and other eukaryotes began to use cytoskeletal elements called *microtubules* in the engulfment of extracellular materials, including nutrients, in a process called *endocytosis*.

Although most eukaryotic protoctists are still single-celled, some of them continued the experiment with multicellularity, and refined it considerably beyond the point achieved by the most advanced prokaryotes. In multicellular protoctists we see for the first time communication among cells living in colonies, and we see the very beginnings of a specialization of labor among cells. Some multicellular protoctists came to be among the largest living things on earth. Kelp, for example, is one example of a type of protoctist called algae; some kelps can grow to be thirty or fourty meters long.

Protoctists also gradually accumulated internal organelles to help them survive in an increasingly hostile and oxygen-filled environment, in some cases through a process called *endosymbiosis*. It seems that certain bacteria began to find the interiors of the larger, more advanced protoctists a reasonable place to live and raise a family. They became parasites. Like most successful parasites, these bacteria gave as good as they took. For example, some bacteria had apparently developed defenses against oxygen; a few even developed means not just to neutralize oxygen but to use it to produce energy. This must have impressed certain protoctists, who apparently had taken in some of these oxygen-breathing bacteria by endocytosis. Instead of digesting them for food, they converted them into permanent parts of the protist cell. This may not have been what the bacteria had in mind, but the experiment turned out to be so successful that eventually all eukaryotic cells acquired similar intracellular parasites. We find these bacterial "living fossils" in our own cells to this day: they are the mitochondria, the energy-producing organelles we met in Chapter One. In animals, only the mitochondria, of all the organelles inside a cell, have their own DNA, showing their independent biological origin. Moreover, this DNA is single-stranded and usually circular, is not associated with histones, and contains genes that are distinctly prokaryotic rather than eukaryotic in structure.

Although bigger may, in many ways, be better for a cell, there is a limit to how large a single cell can grow. A cell's volume grows as the *cube* of its radius (as does its appetite!) while the area of its surface membrane — the site where nutrients enter and wastes are expelled — increases only as the *square* of the radius. At some point in a cell's expansion, its surface area will simply become too small to service the huge internal volume of the cell. Moreover, the need for various internally produced molecules to operate the cell, which are encoded by DNA, also increases dramatically with size. A single set of internal instructions — even a diploid set — rapidly becomes insufficient. Cells solved this problem in a variety of ways in their rush to grow larger. Some simply made more copies of their basic diploid chromosome set, becoming *polyploid*. A few cells became *multinucleate*, incorporating several nuclei, each with only one diploid chromosome set, into a single large cell. But by far the most successful strategy was to become multicellular; all life forms beyond the protoctists, with the exception of a few of the fungi, became completely and permanently multicellular.

It was somewhere along the pathway from the monerans to the protoctists, about a billion years ago, that death as we know it — death as an inescapable consequence of life — first made its appearance. We will call this form of death *programmed death*, to distinguish it from *accidental death* — death caused by things like extreme heat or cold, starvation, physical destruction, or chemical damage. We will also explore the possible relation of this programmed death of the *organism* with the phenomenon of programmed *cell* death discussed in the last chapter. (These terms usually mean something slightly different to the scientists who study them, but that may be too narrow a view.) But first we will look at the very beginnings of programmed organismal death as monerans evolved into eukaryotic protists, for it was also along this pathway, and at about the same time programmed death appeared, that single-cell organisms first began to experiment with sex in connection with reproduction.

The earliest single-cell monerans reproduced asexually in a simple process called *fission*. In this mode of reproduction, a given cell autonomously replicates its own DNA and then divides into two perfectly coequal clones of itself, each clonal offspring receiving one copy of the DNA. These cells mature and each produces two healthy, coequal clones in turn. *Thus the organism — the single cell — never truly dies.* After all, where is the body? Can there be death in the absence of a corpse? These cells are in effect immortal. If a single asexually reproducing bacterium could be protected from predators and supplied with adequate food and space to grow, it would continue clonal expansion through its progeny indefinitely. We would likely never find a single dead cell in such a culture. Of course, in real life, individual cells cannot go on clonally expanding forever. Eventually they would deplete the available resources necessary to sustain life, and they would — accidentally — die.[1]

With only a handful of exceptions, single-cell organisms reproducing exclusively by simple fission lack one feature

---

[1] An average bacterium is about one cubic micrometer in volume and can divide by fission in as little as thirty minutes. Simple arithmetic shows that sixty to seventy cell divisions later — less than two days after starting to divide — the progeny of a single bacterium, if they all survived, would be roughly equal in biomass to all of the human beings presently on earth.

that ultimately brings death to all single cells that have sex, and to all multicellular organisms, including human beings: *senescence*, the gradual, programmed aging of cells and the organisms they make up, independently of events in the environment. Accidental cell death was around from the very first appearance of anything we would call life. Death of the organism through senescence—programmed death—makes its appearance in evolution at about the same time that sexual reproduction appears. Both sex and programmed death began when the vast majority of organisms were still single cells.

It is important to understand that from a biological point of view, "sex" and "reproduction" are two entirely unrelated phenomena. Sex refers only to the exchange or comingling of all or part of the genetic information—DNA—between two members of the same species. Reproduction is simply that—the reproduction of additional copies of a given cell. "Sexual reproduction," then, means the exchange of genetic information in combination with cellular reproduction.

Sexual reproduction has always been something of a puzzle for biologists, especially for evolutionary biologists. From many points of view, sex is a rather wasteful way to reproduce. In fission, one cell gives rise to two; one set of genes is duplicated in a simple, efficient and relatively low-cost operation, and two new individuals result. In sexual reproduction, two cells (or two multicellular organisms) must find each other, determine that they are right for each other, have sex, and then reproduce. Moreover, since most cells engaging in and resulting from sexual reproduction are diploid, two sets of DNA must be reproduced in each cell instead of only one. Thus sexual reproduction requires a great deal more time and energy than simple fission to achieve the same end result. A cell reproducing by simple fission ought to be able to outgrow and outmaneuver a cell reproducing sexually. Yet once sex appeared, it rapidly became the dominant form of reproduction among all subsequent life forms. Why is this so?

Let's take a closer look at sexual reproduction in life forms similar to those in which it must have first been tried during evolution. For single-cell eukaryotes reproducing sexually, sex may involve nothing more than two cells sticking together and swapping portions of their chromosomal inheritance — their DNA — in a process called *conjugation*[1]. After conjugation the cells separate, and then each undergoes fission into two cells that each receive identical copies of slightly scrambled and recombined chromosomes. The act of conjugation — the sex part of the process — starts with two cells and ends with two cells; no additional cells have been created. For single-cell eukaryotes, only when conjugation is followed by independent fission of the participating

[1] It should be noted that a process termed *conjugation* has been described for prokaryotic organisms as well. Although bearing some similarities to conjugation in eukaryotes, the process in prokaryotes is fundamentally different, and not obviously related evolutionarily to conjugation in eukaryotes as described here.

cells can we say that sexual reproduction has taken place. The most important consequence of this act is that the new cells produced as a result of conjugation, and the offspring they subsequently produce by fission, are *genetically different from the parent cells prior to conjugation*. This is quite unlike the process of fission in asexual reproduction, where the parent and daughter cells are usually genetically identical. This difference almost certainly underlies the great advantage of sex, but in the evolutionary line leading to human beings it also contains the seed of a serious problem — a problem that nature ultimately solved by inventing obligatory senescence and death.

This basic process of mixing and exchanging genetic information — sex — is used in connection with reproduction today by some prokaryotes, by the majority of single-cell protoctists and fungi, and in one form or another by most multicellular organisms. There are many theories concerning the origin and evolutionary driving forces behind sex. Whatever else may be said about it, sex unquestionably enhances genetic variation, which is one way species are able to adapt to a changing environment. Different combinations of genetic information generated during mating allow individual offspring to adapt rapidly, albeit with differing efficiencies, to environmental change; those individuals who successfully adapt are selected for survival and further reproduction.

A second major advantage of sex is that it allows for the repair or elimination of genetic mistakes — the mutations that creep into individual genes, causing them to lose their function, whether that function is coding for a particular protein, or regulating some cellular response. Genes are written into the DNA of chromosomes. Over time, different individuals will have accumulated different combinations of defective genes. During sexual reproduction, the chromosomes (and thus the genes) are randomly mixed and redistributed to offspring. Since sex in eukaryotes takes place mostly between diploid individuals, it is unlikely that two individuals having sex will have accumulated exactly the same mutations; mutations in a given gene of one sex partner will most often be compensated in the offspring by the unaltered copy of this gene from the other partner. Moreover, if by chance an offspring should inherit two copies of a mutated gene, that individual may not survive to reproductive age, thus reducing the number of "bad" copies of that gene floating around in the population. Such individuals die (by what we would describe as accidental cell death), and their defective genes die with them.

While sex in the sense of transfer of DNA between cells actually made an appearance among some of the prokaryotes, and while sexual reproduction took several forms among different eukaryotic species, the process that developed in single-cell eukaryotes like the *Paramecium* is most likely ancestral to the way we ourselves use sex. It also demonstrates quite clearly how the coming of sex eventually led to compulsory, programmed death in that line of evolution leading to multicellular organisms like us.

Paramecia are one of a large number of species of ciliated organisms found in freshwater ponds throughout the world. Although single-celled, these protoctists are enormous — easily a million times larger in volume than most bacteria. They use hairlike cilia beating in controlled synchrony to move about, a great advantage for a single cell in terms of escaping danger or finding food. They have even developed a specialized membrane region at one end of the cell that serves as a mouth, and a similar region at the other end that serves as an anus for discharging wastes. Like all eukaryotes, paramecia keep their DNA in a nucleus, of which, importantly, they have more than one kind.

Beginning in the late nineteenth century, biologists interested in cell growth and cell death began to suspect that there might be a fundamental difference between single bacterial cells and the larger single-cell protists like paramecia. Whereas bacterial growth did seem to be limited only by food and space, several investigators thought they observed undeniable signs of senescence— obligatory, time-dependent degeneration and death — in some of the protoctists. For example, if a single paramecium is placed in a glass culture dish (*in vitro*) with unlimited food and space to grow, it will begin clonal expansion by simple fission just as asexually reproducing bacteria do. But before too long the rate of clonal expansion among the progeny of this single paramecium will begin to slow. If the cells continue to reproduce exclusively by fission, this slowing process will continue, and after a total of about 200 cell divisions, the clonal progeny will stop dividing and die. However, if somewhere along the way some of the progeny conjugate — have sex — *their senescence clock is reset.* The clonal progeny of those cells that have sex resume a rapid rate of growth and expansion by vigorous asexual fission. The progeny of their siblings and cousins that fail to conjugate continue to senesce and eventually die.

In ciliates reproducing sexually, the rejuvenated progeny produced by recent conjugation seem to go through a maturation period similar to prepubescence in animals in that they cannot themselves have sex. But they also start life with a predetermined life span. Their senescence clock begins to tick, and they must themselves at some point abandon simple fission and reproduce sexually if they are to reset their own clock and survive through *their* progeny. Immortality for bacteria reproducing only by fission is granted automatically; immortality for everyone else depends on having sex.[1]

Why should the introduction of sex into reproduction have been accompanied by the advent of senescence and programmed death? One reason is related to a problem pointed out earlier: the progeny of cells reproducing sexually are genetically different from the parent cells. A second reason may be related to a feature we see for the first time in paramecia

[1] It is worth noting here that the rejuvenation process in paramecia and other ciliates is tied more to the reproductive act itself than to the exchange of DNA per se, which is the strict definition of sex. Some ciliates are capable of self-fertilization, in which the DNA of a single cell is reshuffled and rearranged (*autogamy*). Although no new DNA is introduced, the cell is nonetheless rejuvenated in the same way as a cell having true sex.

and other protists reproducing by sexual means: *the segregation of DNA to be used for reproductive purposes (conjugation) from the DNA used to direct the day-to-day operation of the cell.* Each type of DNA is kept in a separate nucleus. The *macronucleus* houses the DNA used to direct the production of messages to be sent out to the cell's ribosomes for conversion into proteins needed by the cell to conduct its daily business of eating, respiring, moving about, and so on. The chromosomes, and even selected fragments of chromosomes, are often replicated hundreds of times over in the macronucleus to generate enough DNA to run these very large cells.

Paramecia also have a smaller, relatively inactive *micronucleus*, which contains a single diploid set of chromosomes. Each chromosome in the micronucleus wears its histone coat rather tightly, and for most of the life cycle of the cell lies unread and unused. Only when the cell is about to divide does the micronucleus become active. When paramecia reproduce asexually, both the macronucleus and the micronucleus divide their DNA in two, providing each offspring with a new copy of each type of nucleus. The macronucleus simply divides whatever DNA it has on hand into two pots, one for each offspring. During its brief awakening at the time of cell division, the micronucleus makes a complete and faithful replica of its DNA; one offspring receives the original copy, and one the replica. The newly minted micronucleus goes immediately into hibernation, as in the parent cell, and the macronucleus resumes the day-to-day operation of the new offspring.

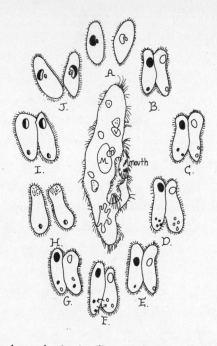

Figure 5. Sexual reproduction in ciliates. A. Two genetically different
ciliates, each with a macronucleus (**M**) and a micronucleus (**m**). **B.** The two
ciliates fuse in the first step of conjugation, and the macronuclei and
micronuclei move to opposite ends of the cell. **C.** Each micronucleus
divides once by meiosis, and **D.** the daughter micronuclei each divide
again, to produce four haploid micronuclei. **E.** Three of the four haploid
micronuclei disappear. **F.** The remaining micronucleus divides once more,
to produce two identical micronuclei, and then the two conjugants
exchange one micronucleus (**G**). **H.** The two haploid micronuclei fuse to
produce a single diploid micronucleus. **I.** The new micronuclei each direct
the production of a new macronucleus; the old macronucleus begins to
disintegrate. **J.** The two ciliates disengage, and the nuclei assume their
starting positions in the cell. The exconjugates are now genetically identical
to one another, but genetically different from either of the two starting
cells. Each will go on to produce genetically identical daughters by simple
fission.

But if the cell finds another cell interested in conjugation (see accompanying Figure), then the micronuclei play an additional role. The two conjugation partners prepare to combine and exchange portions of their micronuclear DNA. This involves an important step that is a feature of sexual reproduction in all eukaryotic cells. The micronuclei are diploid; in sexual reproduction, two conjugating paramecia are going to exchange micronuclei and fuse them, mixing their chromosomes together. This would result in a micronucleus with four sets of chromosomes rather than two. These conjugated micronuclei would be *tetraploid*, and subsequent mating would produce micronuclei with eight chromosomes, then sixteen, and so on ad infinitum. So before the micronuclei are exchanged between two conjugating cells they must undergo *meiosis*. In meiosis, each micronucleus divides into two micronuclei with only a single chromosome set; they become haploid. Like their *monoploid* moneran ancestors that reproduced by simple fission, they now have only a single copy of each chromosome. Each of these new micronuclei assembles its haploid chromosome set by randomly choosing among the maternal and paternal member of each chromosome pair in the parent micronucleus from which it arises. Thus the micronucleus at this stage is already different from the micronucleus from which it is derived.

Once meiosis is complete, the two haploid micronuclei in paramecia divide again, producing four haploid micronuclei. One of these is randomly selected for use in the conjugation process; the others are destroyed. The selected micronuclei divide one last time, and then the two conjugating cells exchange one of their haploid micronuclei. Each cell immediately fuses its mixed pair into a single, diploid micronucleus. These diploid micronuclei define the new individuals resulting from the conjugation process. It is at this stage that compensation for a bad gene in one of the contributing haploid sets can take place. This is also the stage at which a few unfortunate individuals may inherit two defective copies of a gene, and die as a result.

This same process takes place in human reproduction; sperm and ova in their mature state, ready for reproduction, are made haploid through meiosis. After fertilization, the nuclei of the two cells fuse, creating a diploid cell which is the foundation cell of a new human being. But an additional process takes place in ciliates during sexual reproduction. Just as the conjugants are about to separate, the newly created micronuclei undergo additional replications; this time some of the daughter micronuclei continue replicating portions of their DNA to make a new macronucleus. And then the old macronucleus, sitting alone at one end of the cell, begins to degenerate; it dies.

What do ciliated protoctists have to do with human beings? What can they tell us about our own death? A great deal. For it is only in connection with sexual reproduction seen in protoctists like paramecia that we encounter for the first time the *generation of DNA that is not transmitted to the next generation*. This segregation of DNA into two compartments (the macronucleus and the micronucleus) never happens in bacteria or even higher organisms reproducing asexually. And what becomes of this excess DNA not used for reproduction? It is destroyed. In fact, one could very well make the case that *it is in the programmed death of the macronuclei of early eukaryotes like paramecia that our own corporeal deaths are foreshadowed.* The micronuclear DNA is sequestered and protected during the life cycle of these cells. It is used solely for recombination with the DNA from another individual in sexual reproduction. Copies of this recombined, biparental DNA will be passed to the next generation of cells. The rest of the DNA, partitioned off in the macronucleus, is now superfluous. It cannot be passed on to the next generation, because the next generation will be defined genetically by the newly recombined micronuclear DNA. What would the next generation do with this used, genetically different DNA? Moreover, this DNA will have accumulated potentially harmful mutations in the course of previous rounds of asexual reproduction. These mutations are not compensated for or corrected by meiotic sex. The new micronucleus must create a new macronucleus, with new DNA, to run the new cells. Otherwise, what was the point of recombining the micronuclear DNA? If we follow the fate of the old macronuclei forward from these early eukaryotes, we will see all too clearly why we ourselves must one day die.

Admittedly, not all protoctists incorporating sex into their reproductive activities went through the stage represented by the ciliates; not all of them began the process of sequestration of reproductive DNA within the context of a single cell. But very shortly after ciliates segregated DNA within themselves, this segregation and protection of DNA destined for reproductive use would be formalized for all time when some of the protoctists (and most subsequent life forms) became permanently multicellular. As we have seen, multicellularity was explored by bacteria, possibly even before eukaryotes ever appeared. But multicellularity was taken to a completely new level by protoctists and subsequent life forms. And the evolutionary line passing through this new type of multicellularity leads directly to human beings.

Multicellularity has many advantages, among which is certainly size. Multicellularity solves many of the problems mentioned earlier that arise in connection with larger individual cells — too much volume for too little surface, and not enough DNA to direct the operation of very large cells. In multicellular animals, large size as an organism can be achieved with rather ordinary-sized cells. But the greatest advantage of all, as noted earlier, is the ability to assign specific biological functions to different cell types.

Once eukaryotes became multicellular, not only would reproductive DNA be kept in separate nuclei; it would be sequestered in just a few special cells in the body, which in humans and other animals are called *germ cells*. Like micronuclei, germ cells have only one function: the transmission of DNA from one generation to the next via sexual reproduction. The rest of the cells in the body — the *somatic cells* — receive identical sets of chromosomal DNA, but they use this DNA only to carry out the body's workaday, nonreproductive functions. The somatic cells in our bodies divide only by simple fission. They do not exchange or recombine DNA with one another—they do not undergo meiosis or have sex. *The only purpose of somatic cells, from nature's point of view, is to optimize the survival and function of the true guardians of the DNA, the germ cells.* The DNA passed forward to the next generation in germ cells will have been changed, through mixing and recombination with the DNA of another. Many of its mistakes will have been corrected or otherwise compensated for. But like the macronuclear DNA of paramecia, unrecombined somatic cell DNA becomes not just redundant, but irrelevant. It could also become dangerous. Since it is not subjected to random recombination and segregation like germ cell DNA, it would continue to harbor the accumulated mutations of untold numbers of generations, until it became genetic garbage unable to properly run a cell.

Prior to protoctists like paramecia and their relatives, the somatic DNA *was* the germline DNA. Prior to multicellular animals, the somatic cell *was* the germ cell. The haploid germ cells seen in animals like us are in a very real sense the heirs of the micronuclei, and the lineal descendants of the early monerans and the asexual protoctists; only the germ cells retain the potential for immortality. At some point in the life cycle of the individual, they may leave the body proper, combine with other germ cells, and continue dividing to produce progeny that divide and generate yet another multicellular organism with another set of germ cells. When this happens, the germ-cell senescence clock is reset, just as it was after conjugation in paramecia. The rest of the cells of the body, the somatic cells, are condemned — programmed — to senesce and die. Once they have carried out their task of ensuring the survival of the germ cells, they and their excess DNA are no longer needed.

Our own bodies are no different. All of our somatic cells will get old, and they will die. Sex can save our germ cells, but it cannot save *us*. With paramecia and other single-cell eukaryotes, the organism can be rejuvenated by sex because the organism and the germ cell are synonymous. Once the organismal (somatic) DNA was partitioned off from the reproductive (germline) DNA, sex may have become mandatory — even enjoyable — but not rejuvenating. Not for somatic cells. Not for us.

The drive toward ever-increasing size, and eventually multicellularity, led to the creation of extra-germinal (somatic) DNA. The advent of sex in reproduction made it necessary to destroy the somatic DNA at the end of each generation. We do not know which of these two events came first, but we do know that the creation and segregation of nonreproductive DNA never occurred in cells reproducing asexually. Death may not be necessary for life, but programmed death is apparently necessary to realize the full biological advantage of sex as a part of reproduction. Not all cells that experimented with sex during evolution created somatic DNA, but the protoctists on the evolutionary path leading to humans and other animals did. In part this was a response to the need for more DNA to direct the operation of ever larger individuals. But once this tendency was combined with sex, death was the inevitable outcome.

Just as with asexual single-cell organisms, of course, many of our somatic cells will die before their time for reasons other than senescence, like the heart cell in Chapter One that succumbed to ischemia. Others will die from infection or poisoning, or will be killed by our own immune systems in the process of clearing an infection. If we lose a limb in an accident, the cells of that lost limb will die within minutes of its disconnection from the body. But all the somatic cells in our bodies that somehow manage to escape accidental death or death caused by disease will still one day die of "natural causes" — of senescence. We will explore the way in which this happens in the next chapter.

4

From Sex to Death: The Puzzle of Senescence

O how shall summer's honey breath hold out
Against the wreckful siege of battering days?
                                          —William Shakespeare

Senescence is the clock that marks our passage through life; if we escape all other forms of death, when this clock runs down we die. But if the death of an organism can ultimately be accounted for by the deaths of individual cells, then what does senescence mean at the cellular level? How does a cell grow old? When a cell dies as a result of senescence, how does it die? We have said that among all the cells in the body, only germ cells retain the potential for immortality; only germ cells are able to reset the senescence clock. How do they do this? These are among the most important questions in the study of aging, and are of great interest to cell and molecular biologists as well.

There are many theories about the biological mechanisms underlying senescence, but they all fit into one of two categories. One school of thought (the "catastrophists") believes that as cells stop dividing, at about the time an organism reaches physical maturity, synthesis of the various molecules of which they are made slows greatly, or even stops. In cells that divide repeatedly, like single-cell organisms reproducing by fission, or like the cells making up a growing embryo or a young child, these molecules are constantly renewed and the component parts of each cell remain young and healthy. But in a mature organism, one that has reached the limits of its growth, this pattern of ongoing cellular replacement and renewal comes to an end. Over time, the molecules simply wear out and are no longer able to do their job. Eventually, if enough copies of critical molecules lose their function, the result for the cell is catastrophic; its structure begins to erode, or it is not be able to carry out vital tasks, such as energy production or movement; perhaps some of the critical membrane pumps break down.

The second school of thought (the "genetic programmers") believes the answer lies deeper. Given that the early stages of growth and development, which are also time-dependent and irreversible, are under strict genetic direction, why would we assume that the continuation of these stages into decay and death is not? Moreover, if it were simply a matter of parts wearing out, what would explain the tremendous differences in life span of multicellular organisms that are all made of exactly the same basic molecular materials? There is no real difference, after all, between the proteins in a mouse and the proteins in a human. Why does one animal live three years, and the other eighty? Perhaps in the world of multicellular organisms death is too important to be left to chance. Perhaps death is genetically programmed, in the same way eye color or cholesterol level is.

As is so often the case in biology, when two schools of thought each have sound and convincing arguments to put forward, both are probably right. The two viewpoints just presented are not mutually exclusive. The molecule we would have to worry most about in terms of wear is very likely DNA — the repository of all genetic control. DNA is constantly threatened with mutation, either internally, from mistakes made during the replication of DNA that accompanies cell division, or from external agents such as chemicals or radiation. Haploid DNA in germ cells undergoing meiosis is readily accessible and easy to repair, and germ cells are literally loaded with the equipment to do it: DNA-repair enzymes. Somatic cells have a much shorter supply of these enzymes, especially as they get older, and so their DNA is much more difficult to repair. As a result, somatic cells gradually begin to accumulate uncorrected mutations. To the extent that this mutant DNA is called upon to make more proteins needed by the cell, it might well make faulty proteins that would compound the problem of existing normal proteins wearing out. The rate at which the accumulation of lethal mutations happens in different organisms could be dictated by the efficiency of the DNA repair enzymes or by their rate of disappearance from somatic cells, either of which could be genetically programmed.

There is reasonably good evidence that senescence in ciliated protozoa that fail to have sex is in fact largely caused by the accumulation of mutations in macronuclear DNA. If macronuclear (somatic) DNA were passed on from generation to generation, without the corrective repairs available to germ-cell DNA, this DNA would eventually mutate into something completely useless. On the other hand, micronuclei (like germ cells) have high levels of DNA repair enzymes, which are themselves encoded by genes in the DNA. Thus in those protists that have sex, it is clear that limited life span is a genetically determined property of the organism, and proceeds at a specified pace independently of conditions outside the cell.

There is also evidence in favor of the genetic determination of longevity in humans. For example, studies with twins show that genetically identical twins on average die thirty-six months apart; their lifespans are very similar. By comparison, fraternal twins die seventy-five months apart, and randomly selected siblings have an average time between deaths of 106 months. The closer two individuals are genetically, the closer their life spans. There are also genetically based diseases in which the aging process is greatly accelerated.

One such disease is the rare but tragic Hutchison-Gilford *progeria syndrome*, in which children undergo the entire human aging process, through death, in about fifteen years. The first changes appear in the affected children's skin, which in the first year or two of life becomes wrinkled, thin, and parchment-like, almost translucent. Their faces begin to look old, with delicate blue veins criss-crossing their foreheads. A few years later, their hair begins to fall out; what is left soon turns gray. These children never thrive; they are chronically underweight and shorter than normal. They lose their body fat, just as old people do. They have skeletal problems, and they become wizened and bent before their time. They rarely enter puberty, seeming to progress directly into old age. Frail and shriveled, they usually die of cardiovascular disease or stroke before the end of their second decade of life. Everyone agrees that this phenomenon of premature aging is genetic in origin, although the precise nature of the gene defect and its mode of transmission are unknown. Obviously this gene, whatever it is, is critically involved in the aging process.

A dramatic demonstration of senescence at the cellular level, one that also points to genetic control of aging, can be seen in isolated human cells taken from healthy individuals and cultured outside the body. Human connective tissue can be dissociated into single cells by mild digestion with certain enzymes. The resulting single cells are mostly of a type called *fibroblasts* — the "worker cells" in Chapter One that helped make scar tissue in our patient's damaged heart. Fibroblasts

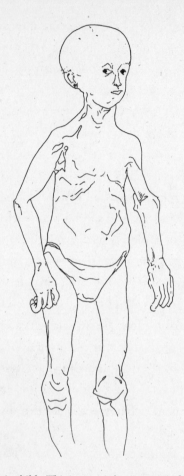

Figure 6. Progeric child. This is an artist's composite of a male child, age approximately thirteen, showing the classic signs of the Hutchison-Gilford syndrome, or progeria. The head is larger than normal, with rather small facial features including a slightly beaked nose and a receding chin. There is extensive hair loss, and fine, bluish veins on the scalp and forehead. The skin over most of the body is sagging and wrinkled. Arthritis is evident in many of the joints, particularly the knees and elbows.

are not unlike many single-cell organisms; although they are many times larger, each cell is nonetheless a discrete, self-contained entity with but a single diploid set of the DNA blueprints representing the person it was taken from. If put into culture with an unlimited supply of nutrients, and kept at body temperature in a humid environment with an appropriate balance of oxygen, nitrogen, and CO_2, fibroblasts immediately start dividing, reproducing themselves asexually exactly as bacteria or nonconjugating paramecia would. Each division produces two identical daughter cells. Though such a large and complicated cell takes much longer to divide than a bacterium, or even a paramecium, fibroblasts nonetheless enter into a state of smooth and essentially uninterrupted proliferation.

The difference between our cells and single-cell prokaryotes like bacteria becomes apparent only after a few weeks in culture. As long as we keep replenishing the nutrients in bacterial cultures and siphon off the excess bacteria to avoid crowding, the bacterial cell division rate never changes. But human fibroblasts, no matter what we do to them, no matter how often we change the medium or enrich it, eventually begin to slow down. It takes longer and longer for each cell to divide into two daughter cells. Finally, like asexual paramecia, fibroblasts simply stop dividing, and nothing we can do will make them start up again. It is as if they had been following some internal program for proliferation, and had come to the end of that program. A week or so after the final round of cell division, they curl up and die.

This phenomenon was investigated in great detail by the renowned cell biologist Leonard Hayflick in the 1970s and 1980s. He found that fibroblasts isolated from a human fetus could undergo, on average, a total of fifty rounds of cell division outside the body before finally stopping. Fibroblasts isolated from a middle-aged person might divide twenty or thirty times before slowing down, degenerating and dying. Cells taken from a very old person divided only a dozen or so times, and only very slowly. Remarkably, if fetal fibroblasts were stopped after twenty doublings, frozen and stored in liquid nitrogen (for up to twelve years), and then returned to culture, they would divide just about thirty more times before slowing and dying. These cells appeared to continue on toward completion of the program begun years earlier, as if nothing had happened, as if time itself had stopped.

These experiments with human cells in culture have been repeated using fibroblasts isolated from children afflicted with progeria. Tiny pieces of skin were removed from three such children, from their parents, and from normal children of the same age and sex as the progeric children. The fibroblasts from the skin of the healthy children took off rapidly in culture, dividing frequently and expanding at a great rate. The cells taken from the parents grew well for awhile and then slowed down, lagging far behind the cells from the healthy children. The fibroblasts from the progeric children grew even more slowly than those from the parents. They divided a few times, stopped, and then died, foreshadowing the early death awaiting the children from whom they came.

All these studies have contributed to the notion that animal cells have a built-in program limiting their lives to a predetermined span. In the case of human fibroblasts this program can be observed directly in the number of doublings they undergo before they die. In other cells that do not normally divide when placed in culture — muscle cells, for example, or nerve cells — the gradual loss of other functions may be more important, but the principle is the same.

How do cells that reach the limit of their ability to divide and function die? Again we can get a glimpse of what may be involved by looking at experiments carried out in one of the protists, in this case a single-cell ciliated cousin of the paramecia called *Tetrahymena*. Like paramecia, tetrahymena have two nuclei — a macronucleus, containing DNA used to run the cell, and a micronucleus, housing DNA used solely for reproductive purposes. And like paramecia, tetrahymena senesce in the absence of sex. When tetrahymena conjugate and are preparing to produce daughter cells, the micronuclei divide to produce new micronuclei, which then go on to make new macronuclei. The old macronuclei degenerate in a familiar way: the chromosomes begin to condense, and the DNA begins to fragment into tiny, unreadable pieces; the membrane of the old macronucleus breaks down, and soon the entire nucleus disappears. The degraded materials are used by the cell as food. This process is very specific for the old macronucleus; the new macronuclei (and the micronuclei) are untouched. Somehow the old macronucleus has been told to initiate a program of self-destruction, and as far as can be discerned, this program produces exactly the same nuclear destruction that accompanies apoptosis in higher organisms. Recent studies suggest that at least part of the death program is encoded by genes in the old macronucleus itself. As we saw for other cells, one of the last acts of the old macronucleus is to transcribe genes governing its own death.

The lesson of the protoctists is thus that the repair of somatic (nonreproductive) DNA is troublesome, expensive, and in the end just not worth it. Besides, as we saw earlier, somatic DNA must be changed to reflect the composition of the new, sexually produced reproductive DNA. This would be difficult enough in the macronuclear DNA of single-cell ciliates; in multicellular animals, it is out of the question. It is easier simply to destroy the old somatic DNA and start over. If that somatic DNA is in separate cells, the cells die too. Unfortunately, those cells are us.

So in fact, programmed cell death is even older than the nematode *C. elegans* described in an earlier chapter; it goes all the way back to the earliest sexually reproducing single cells. The example of the protoctists raises a startling possibility for multicellular animals: is death at the end of senescence ultimately the result of cell suicide? Cell suicide — apoptosis — as we ordinarily think of it in multicellular organisms might seem at first glance to have little to do with senescence. Animal cells commit suicide most commonly during the active years of an animal's life, particularly during embryonic stages in the shaping of an emerging life form. Cell suicide may come into play in the middle years as a defense against radiation damage, or during the course of an immune response. Senescence is seen only at the very end of an animal's life, as a process leading up to the destruction of unneeded somatic cells. But are apoptosis and death as the endpoint of senescence really different? Both are strongly believed to be genetically programmed and regulated. Both types of death seem to have arisen at exactly the same time in evolution. In tetrahymena and paramecia, at least, the function of macronuclear "suicide" is related to sexual reproduction: the need to dispose of DNA not transmitted to the next generation. But isn't that exactly what the elimination of unneeded somatic cells is all about?

One way to gain some insight into the relationship between senescence and programmed cell death is to observe what happens to somatic cells when they come to the end of the "Hayflick limit." When we monitor fibroblasts during the final stages of their lives in culture, we find that over time they lose their ability to divide, and eventually simply sit there, waiting for death. When death finally comes, it exhibits all the classic signs of apoptosis — membrane blistering, nuclear disintegration, the formation of apoptotic bodies and so forth. The fibroblasts in fact die by what we would have called suicide in younger cells. Is the death of cells at the end of the life of a multicellular animal also suicide? Does programmed cell death determine the lifespan of a cell, and thus of an individual?

It is perhaps unlikely that all of the features we associate with senescence — for example, the tissue degeneration seen in old people or in progeric children — are part of the same program as apoptosis *per se*. But it is entirely likely that at some point, as cells gradually lose their ability to function as a result of the action of genes controlling aging, they realize that the game is up and that it is time to step aside. It is time to do what all somatic cells must do once they have finished their task. That task is simply to assure the survival of the guardians of the DNA, the germ cells. When that task is done, they — and we — must die. Suicide may be as graceful a way as any.

We pointed out earlier that germ cells are like asexually reproducing single cells, and like the micronuclei of some of the protoctists, in that they are potentially immortal. Part of the reason for this may be that they are able to purge harmful mutations from their DNA. We ourselves descend from these immortal cells; how do we become mortal? What happens to a potentially immortal haploid germ cell after fertilization, when it fuses with another potentially immortal haploid germ cell to become a diploid *zygote*; what is the "mortality state" of this founding cell as it embarks on the construction of a completely new individual who will be mortal? When this single cell divides into two, and then four and then eight, is the property of immortality lost, or shared equally among all the progeny, or is it perhaps hoarded and

nurtured by just one or a few of these cells that remain immortal, eventually giving rise in a direct lineal fashion to the
new individual's germ cells?

The answer seems to be that at least for awhile, all the
cells of a growing human embryo retain the property of immortality. And then, as the embryo begins to undergo the series of cellular specializations known as *differentiation*, which
eventually give rise to individual organs and tissues, *all* of the
cells of the embryo become mortal; they begin the process of
senescence, albeit without any immediately visible impact on
the developing individual. But the pathway leading to death
is set in motion before the embryo even assumes human
form, and the immortality of the germ cells is conditional: it
will have to be restored somehow at a later stage.

The immortality of the cells of the early embryo can be
seen in the laboratory through the study of *embryonal stem
cells* (ES cells). These cells, normally prepared from mice, have
been used by many laboratories around the world to produce
genetically altered strains of mice — most commonly, mice
that are missing a specific gene. The close study of such mice
can reveal a great deal about the role of the gene in question.

ES cells are obtained by removing the developing mouse
embryo from the oviduct after fertilization but before im-
plantation in the uterine wall. At this stage the embryo has
perhaps sixty to100 cells. Scientists have been able to disso-
ciate this ball of cells, called a *blastocyst*, into single cells and
grow them in vitro. By carefully manipulating the culture environment, it is possible to prevent the ES cells from pro-

ceeding down the pathway to differentiation, and yet allow them to continue dividing. ES cells have been maintained this way through many dozens of passages in vitro, enabling just a few harvested cells to expand into many millions.

If these cells are reintroduced into a freshly isolated blastocyst, they will participate in the creation of a perfectly normal new individual. They will contribute to any and all parts of that new individual, including the germ cells, showing that they have not lost any of their capacity for generation; in the language of developmental biologists, they have remained *totipotent*. By selective mating of the offspring of mice produced in this fashion, it is possible to produce a mouse that is genetically identical to the donor of the ES cells, a mouse which will in turn produce offspring, including germ cells, of the donor type. Thus we can conclude that the cells from the early embryo — the ES cells — were both totipotent and immortal. But if we harvest cells from later stages of embryonic development, this type of experiment will not work; it is clear that cells beyond the ES stage (the blastocyst stage) have all lost their capacity for producing a new individual, and that they are all mortal. They have begun the process of senescence, and although they may grow for a longer time in vitro than cells from an adult, they will in fact all die.

The notion that mortality is a specific, genetically controlled program is an important one. From an evolutionary point of view, the appearance of senescence and death in somatic cells and the resulting destruction of somatic DNA was a "gain of function" event; these properties did not exist in cells for the first billion or so years that life existed. Once senescence and programmed death appeared in certain organisms, they—and the genes underwriting them—became "fixed" because they were advantageous. Senescence and death have to be actively worked toward; they do not just happen. Somatic cells even have a number of safeguards built in to make sure they do not backslide and try to become immortal. But to whom or to what are these programs advantageous? Who or what profits from our aging and death? The only answer possible, the only conceivable beneficiary, is the DNA being passed from the previous generation to the next generation via the germ cells. And this DNA will carry with it instructions for the senescence and death of the next generation of somatic cells.

Can human somatic cells ever escape the fate programmed in their genes? Can they ever escape the Hayflick limit and become immortal? Consider the extraordinary case of Henrietta Lacks, a thirty-year-old African-American and an apparently healthy mother of four who was diagnosed with cervical cancer in February 1951. Gynecological and breast cancers in young women are relatively rare, and usually difficult to treat successfully. Henrietta Lacks was admitted to the prestigious Johns Hopkins Hospital in Baltimore, where she lived. A small piece of her tumor was removed for study by a pathologist, who confirmed that it was a particularly aggressive form of cancer. She was immediately given radiation treatment in the area of the tumor, and the initial results were encouraging. On several follow-up visits over the next few months, no tumor could be seen by direct visual examination. However, during the summer of 1951 she complained of increasing abdominal and kidney pain. Although cleared from her cervix by the radiation treatments, the tumor had spread to other nearby organs. Despite heroic efforts to save her, Henrietta Lacks died in October of that year, barely eight months after the initial diagnosis. It was indeed an unusually aggressive tumor.

But Henrietta Lacks' story did not end with her death. A portion of her tumor — the slices the pathologist had examined that February — was eventually passed on to a researcher at Johns Hopkins named George Gey, who was interested in how viruses, particularly the polio virus, grow in human tissue. To carry out his studies he needed human cells that could be grown in vitro. Gey was one of the leaders in the emerging field of human cell culture, but his successes so far had been marginal at best. Like others — anticipating Hayflick's later studies — he found that human cells put into culture would grow for awhile and then stop, which made experiments with viruses very difficult. Experiments with a particular cell line would yield tantalizing information about viral growth or other properties, and then the line would disappear. The virus would have to be passed to a new cell line, often from a different tissue source, and almost certainly from a different person. The virus might or might not behave in the same way in the new cells.

Gey took the biopsy samples of Henrietta Lacks' cervical tumor back to his lab and began trying to grow cells from them in vitro. As was his usual practice with human cell samples, he identified them by the first letters of the donor's first and last names: in this case, HeLa. Little did he imagine that this name would survive not only Henrietta Lacks but himself and most of his coworkers.

After a week or so in culture, it became apparent that these were very unusual cells. They grew vigorously, required constant feeding, and had to be thinned out frequently to prevent overcrowding in the culture dishes. Viruses loved to grow in them. Everyone in the lab was delighted. Here at last was a stable human cell line that could be used to study disease-causing viruses. There was some nervousness about the fact that this was a tumor cell line, and thus possibly "not normal." Would the results obtained with such a cell line be applicable to normal cells? Comparisons of information obtained with HeLa cells with previous results in the laboratory apparently reassured the researchers. Investigators interested in other aspects of human cell biology asked for samples of Gey's new HeLa cells; invariably, they seemed to work beautifully. HeLa cells appeared to be ideally suited to the study of a wide range of biomedical questions.

Soon HeLa cells became the most heavily used human cell line in the world. They were distributed to researchers in virtually every country, including (in the interests of detente) various republics of the Soviet Union. HeLa cells were even sent into space aboard the Discoverer 17 satellite. This widespread proliferation eventually led to a minor scientific scandal, and a major research problem. While delighted to use HeLa cells in their studies, many researchers also continued trying to grow other human cell lines from other tissue sources — liver, kidney, or heart, for example. Such cell lines were considered important because of the tissue-specific properties they presumably possessed. But almost every lab also grew HeLa cells, which were so much more vigorous and aggressive than other cell lines that if even one single HeLa cell somehow got into a culture of another cell type, the HeLa cells would quickly take over, crowding the other cells out. This apparently happened quite often. In 1966, Stan Gartler, a geneticist at the University of Washington, tested a number of human cell lines supposedly of different tissue origins, and showed that the majority of them were in fact HeLa cells. This was a major setback for researchers all over the world who had already published, collectively, hundreds of scientific papers based on the presumed properties of heart cells or liver cells. In fact most of the cells they were describing were HeLa cells. The researchers were not amused.

This story illustrates once again the principle stated earlier: death is not an automatic corequisite of life. HeLa cells, and other human tumor cell lines that have since been established, behave exactly like primitive single-cell organisms. They proliferate asexually by simple fission. Give them unlimited food and oxygen, thin them out periodically to prevent overgrowth, and they will live indefinitely. Their "clock" is perpetually reset; they do not senesce and die. Like germ cells, they are potentially immortal. The number of HeLa cells in the world today — all of which still contain within them the DNA hologram describing Henrietta Lacks — is impossible to estimate. If properly fed and cared for, HeLa cells double at least once a day. As of the end of 1994, HeLa cells had been around for over 15,000 days. Each cell put into culture in 1951 could theoretically have produced $2^{15,000}$ progeny. Such figures, even when accidental loss and purposeful destruction of cells through the years is taken into account, defy comprehension. But beyond any shadow of a doubt, the DNA blueprints for creating a Henrietta Lacks are the most plentiful and widely distributed set of such instructions in the world today.

Human tumor cells appear to have reverted to that state of initial grace granted to the first cells on this earth, potential immortality. Unfortunately, unless removed from the body and cultivated in vitro like HeLa cells, they never enjoy their newfound freedom for long; with their voracious appetites and need for space to grow, they eventually kill their hosts, and themselves in the process. An immortal tissue in a mortal body is a recipe for disaster. Scientists have found that cells infected with certain viruses may escape the curse of senescence as well, and enter into perpetual youth and renewal. But these virally transformed cells too are a threat to the health of their hosts, and thus eventually to themselves.

How do cancer cells do it? How do they turn off their senescence clock and avoid ultimate programmed death? The answer intersects in interesting ways with the loss and conditional reacquisition of immortality in germ cells. As we have seen, the early cells of the human embryo retain the immortal property of their germ cell progenitors. At this stage, the growing embryo is in fact growing without any regulation and is dangerously similar to a tumor. But this unregulated growth is quickly brought under control as the embryo begins to differentiate. All of the cells suddenly become mortal; a special subset of these mortal cells will differentiate into germ cells at a slightly later stage, and reacquire immortality. How does this come about?

One of the things that happens during embryonic development and differentiation is the gradual shutting down of the *genome* (the term used to refer to the entire collection of DNA sequences scattered along all of the chromosomes). Germ cells and early embryo cells at or before the stage represented by ES cells appear to have what we might call an "open genome": essentially all of the genes of the genome are open and accessible, ready to contribute to the various structural and functional components of the new individual. As differentiation proceeds, however, the various cells of the embryo begin to shut down entire blocks of genes, committing themselves thereby to become only the types of cells specified by the remaining "open" DNA. Different cells leave different portions of their genomes open; the particular set of genes remaining in the open state is what gives each cell type its unique properties. Developing embryonic cells thus travel along different pathways, but they all proceed from totipotency to *pluripotency* (having a limited potential for further development), and eventually to become single, specific cell types with very limited gene expression and no potential for further or alternate development. A kidney cell, once fully differentiated, no longer has the capacity to become a lung cell; a brain cell can never become a blood cell. All of the genes that were originally present in the open genome are still physically present in every fully differentiated cell, but in each such cell the vast majority of the genome is in a greatly repressed and inaccessible state.

It is during this transition from totipotency to the final differentiated state that the embryonic cells become mortal. The potential for future cell division is sharply curtailed and in most cells will eventually disappear, and the process of senescence begins. Understanding how this takes place, and how it may relate to cancer, is one of the most active areas of contemporary biological research. The current best guess is that in fact the program for limited cell division and senescence is present and operational even in germ cells and early embryo cells, but that it is effectively neutralized by *death repressor genes* whose products interfere with the senescence program and the limits placed on proliferation.

Under this scenario, there would be sets of genes — we'll call them *death genes* — whose expression initiates senescence and loss of the ability to replicate chromosomes, leading eventually to death of the cell. (These would be the genes acquired by certain cell lines in the evolutionary "gain of function" event described earlier.) In open genomes, like those found in germ cells and early embryo cells, the death repressor genes are fully functional, allowing uninhibited cell division and blocking senescence; as long as these genes are expressed, the cells are effectively immortal since the senescence program cannot operate. As the cells begin to differentiate and turn off large blocks of genes, the death repressor genes are among the first to be turned off. The death genes themselves are never fully shut off. Again, as we saw earlier in a different context, death is the default state. Tumor cells appear to have found a way to turn some or all of the death repressor genes back on, or to turn the death genes off, and to a greater or lesser degree mimic germ cells. In fact the majority of tumors show almost no signs of cell specialization; they either have "de-differentiated," or have arisen from a small pool of cells within each tissue that show limited degrees of differentiation.

One of the events contributing to senescence, and ultimately to the activation of programmed cell death, is very likely the accumulation of mutations in somatic-cell DNA, to a point where the DNA begins coding for too many faulty structural or functional proteins. Somatic cells appear to be set up with monitoring equipment that allows them to know when damage to DNA is approaching the critical point; when this point is reached, the death genes become active and the cell is instructed to commit suicide. As pointed out earlier, germ cells and early embryo cells express high levels of DNA-repair enzymes, which would prevent accumulation of DNA mutations in the first place. These enzymes would thus be candidates for products of death repressor genes. (However not all tumors that escape control of cell replication acquire the enhanced levels of repair machinery necessary to keep their DNA and chromosomes normal; many tumors have highly abnormal chromosome structures and mutant genes, as do fibroblasts kept for long periods of time in culture. The truly successful tumors, however, clearly have acquired enough of the germ-cell-like state to keep their DNA and chromosomes in good working order.)

Recently, cancer researchers have begun to focus attention on a structure we encountered in the last chapter — the chromosomal telomeres. Recall that when eukaryotes linearized their previously circular chromosomes, they had to cap the ends with telomeres to keep them from recircularizing, or from sticking to each other end to end. Telomeres are themselves made of DNA, but in a form less sticky than ordinary DNA. When cells divide, the telomeres are not reproduced along with the rest of the DNA in the chromosome; they are added anew at the chromosome tips after each round of cell division, using an enzyme called *telomerase*.

Researchers had noticed that as individuals age, the telomeres at the ends of their chromosomes gradually get shorter. The same thing can be seen in fibroblasts grown in culture; fibroblasts taken from a young person start out with long telomeres, but as the cells age in vitro, the telomeres grow shorter until they almost disappear. Fibroblasts isolated from progeric children have extremely short telomeres. It is believed that as telomeres shorten and disappear, chromosome ends begin to stick together, making chromosomal replication of DNA (and hence cell proliferation) virtually impossible. There had been suggestions that telomere shortening might be related to senescence, but it was not obvious whether it was a cause or an effect. Recently, however, researchers took a closer look at HeLa cells which, as we have seen, have gone through a huge number of cell divisions since they were first put into culture forty-five years ago. To everyone's surprise, HeLa cells have long, healthy-looking telomeres, typical of those found in germ cells and in the cells of very young individuals. They also have very high levels of telomerase. This is a condition found in many long-term human tumor cell lines and again, there is a strong resemblance here with germ cells: both germ cells and early embryo cells have high levels of telomerase activity. ES cells also have very high levels of telomerase, and maintain their telomere length through an indefinite number of cell divisions. But once the early years of growth and increase are over, telomerase activity declines and the telomeres gradually shorten. Telomerase may thus be another good candidate (at least in humans) for one of the death repressor genes.

As for the death genes themselves, a number have been identified, but one of the most interesting continues to be a gene coding for a protein called p53. This gene plays several important roles. For example, when cells have been irradiated to a point where their DNA is damaged, the p53 gene is activated and induces the cell to commit suicide. The p53 gene also prevents cells from proliferating when they shouldn't: it prevents them from becoming cancerous. When normally quiescent cells try to enter into active cell division, the p53 gene is again turned on, and the cells undergo apoptosis. Not surprisingly, mutations in p53 (which render it functionless) are the most common mutation seen in human cancer; cells that have lost p53 often begin to divide without any control. Supporting this clinical observation is the fact that mice that have had the p53 gene knocked out have extremely high spontaneous cancer rates. Finally, p53 appears to be involved in the normal senescence of human cells. When human fibroblasts are placed into culture, as we have seen, they eventually age and die by apoptosis. But if the p53 gene is somehow silenced, senescence is greatly retarded and a large number of the fibroblasts actually become immortalized — they become like HeLa cells.

The correspondence between early embryonic cells and tumor cells has fascinated researchers for many years. A great deal has been learned about the death of cells by studying cells that have somehow evaded it. And it is inescapably true that the death of a human being begins with, and is ultimately entirely explainable by, the death of individual cells. The two deaths cannot be separated from one another; they are the same death, whether we write on the hospital chart that death came from a "heart attack" or "cancer" or simply "old age." Yet as we have seen death is not, a priori, a requirement of life. Somatic cells — and thus the need for compulsory somatic cell death — arose only after DNA began making copies of itself that would be used for purposes other than reproduction. For humans this means that once a reasonable number of our germ cells have been given a chance to impart their reproductive DNA to the next generation, the rest of us — our somatic selves — becomes so much excess baggage. That is the biological origin of senescence and death.

From a human point of view, it is our somatic selves — embedded in which are things like mind, personality, love, will — that we cherish most and that define us, to ourselves and to others. We think of reproduction as only one of many activities we can choose to engage in. Perhaps this is not surprising, since it is a point of view arising in the somatic part of ourselves — in our minds. We have used our minds to invent complex belief systems to explain death. None of these paint a picture of ourselves as excess baggage; none cast us simply as tools for transmitting DNA. Yet when we trace the origin of our death beyond mind and belief, to its true beginnings — the death of individual cells — we come to a rather harsh and unflattering conclusion: the irrelevance, in the grander scheme of the universe, of our somatic selves. No wonder belief so often triumphs over reason.

5

A Hierarchy of Cells: The Definition of Brain Death

The Brain is just the weight of God.
           —Emily Dickinson

Let us turn once again to our patient, who we left in the back of an ambulance speeding toward a hospital emergency room. The paramedics continued to provide oxygen and to monitor his vital functions closely during the ride. Fortunately the morning traffic was still light; they reached the hospital in just under five minutes. When he was brought into the ER, he was still unconscious. His heart was beating regularly, but more slowly than it should. His breathing appeared relatively normal.

Immediately upon his arrival in the hospital, steps were taken to decrease the possibility of further damage to his heart. (All the damage is not done at the moment of the attack; it evolves slowly over a period of several hours.) He was connected to a more sophisticated heart monitor (ECG), and to a hospital source of pure oxygen. A second I.V. line was started to facilitate drug and nutrient delivery. Immediate measures were taken to decrease the demands placed on his heart. He was given clot-inhibiting agents to prevent further blockages in the arteries feeding the heart muscles, and lidocaine to prevent irregular heart rhythms from developing. Blood samples were drawn and sent to the laboratory for analysis to help determine the precise extent of the damage. Dead cells, as we have seen, release their contents into the lymph, which eventually makes its way back into the bloodstream. The presence in the blood of substances normally found only inside healthy cells can tell the clinical chemist which cells have died, and approximately how many and how long ago.

Our patient's slow heartbeat (*bradycardia*) worried the emergency room physicians; atropine and dopamine were introduced through his I.V. to speed up his pulse. He was monitored very closely; this is a risky procedure. On the one hand his doctors wanted to reduce the load on the heart, but the potential danger of a feeble pulse is much greater, because it slows delivery of life-giving oxygen to the tissues, including the brain. He was stable enough after an hour or so to be moved to the coronary intensive care unit. Physiologically he is now relatively stable; but he has not regained consciousness, and this does not bode well.

As we stand here in his room and watch him lying in his bed, he looks quite normal in many ways. His hair is a little matted, and he seems sunken back into the mattress. The events of the past few days have clearly taken a physical toll. But he is breathing well on his own now, and he is warm to the touch. A physician shines a light into his eyes, and his pupils contract normally. He jerks away in response to painful stimuli. When food is delivered through a gastric tube or an i.v. line, he digests it and sends the nutrients out through the bloodstream to his cells and tissues, which efficiently use them in combination with the oxygen his breathing provides. His kidneys process and excrete the wastes produced by his cells.

On the other hand, he still has not opened his eyes. He has no knowledge of the room he is in, or of the people who come and go to minister to him, or indeed of his very existence. He is in a deep coma. The initial laboratory tests, bedside examinations and electroencephalogram (brain wave) patterns suggest the possibility that he may be *apallic*: that the cortex of his brain, which is highly sensitive to oxygen deprivation, may no longer be functioning, but that his brainstem, the less oxygen-sensitive *encephalon*, is still alive. All of the responses described above, such as temperature control and pupillary contraction, are actually involuntary

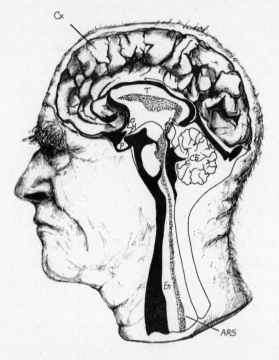

Figure 7. Major anatomical divisions of the brain. The *cerebral cortex* (**Cx**) is the region of the brain housing those functions most associated with "human-ness": thinking, memory, awareness of oneself and of the environment. The *thalamus* (**T**) is like an air traffic control tower; it sorts out, coordinates and integrates the various sensory inputs coming into the brain. The *hypothalamus* (**H**) and *pituitary gland* (**P**) control the brain's hormonal connection with the rest of the body. The *cerebellum* (**Cb**) helps us to keep our sense of balance and physical coordination in relation to the world around us. The brainstem, or *encephalon* (**En**), controls basic body functions (heart rate, breathing) as well as reflexes (pupillary reflex, gagging, swallowing). It also contains the origins of something called the *ascending reticular activating system* (**ARS**), which regulates alertness and wakefulness versus sleeping.

reflexes controlled by the encephalon; they require no high-er brain function or coordination.

If he does not regain consciousness in the next several days — and particularly if he remains comatose — the likeli-hood will increase that he has indeed lost all higher cortical functions. It is possible he may emerge from coma but still not recover consciousness; he could enter a *persistent vegeta-tive state (PVS)*. More tests will have to be done, but it is a pat-tern the nurses and physicians in the coronary ICU have seen all too often. In spite of heroic efforts by his wife and by the responding life-support teams, parts of his brain may simply have been deprived of oxygen for too long a time. The diffi-culty in reestablishing breathing may well have been the crit-ical factor, but it will be impossible to know.

We will return to see how our patient finally fares in Chapter Seven. He will be moved shortly to the general in-tensive care unit, where he will be watched closely over the coming days. A clear diagnosis will at a minimum take sev-eral weeks of observation and testing. At some point the hos-pital staff will have to reach a conclusion about the state of their patient's brain function, and report that conclusion to his wife. In the meantime, we should begin to consider the fine line in cases like this between life and death. It is very complex; it is the point at which the death of a cell and the death of a person would appear to diverge.

At first glance our patient seems very much alive. However, if enough of his brain were to be judged com-pletely nonfunctional he could in fact be pronounced dead, regardless of how lifelike he might seem. But how much of his brain must be dead for such a declaration? What if his brain is not completely dead, but he never regains consciousness? Is he then really alive? How and by whom are such decisions made?

The current medical standard for deciding whether someone is legally alive or dead in cases like this is what is called the *brain-death standard*. The impetus for establishing such a standard arose from a variety of needs created mostly by advancing technology, precisely the type of technology that has kept our patient alive for the past seventy-two hours. Prior to about 1950, declaration of death was relatively uncomplicated, and basically any physician could declare someone dead. The criteria were vague, revolving around the commonsense ideas that someone whose heart is not beating, or who is not breathing, is almost certainly dead. If the physician was slightly off in his or her determination of an exact moment of death, no matter. Anyone without a heartbeat or unable to breathe would surely be dead in a matter of minutes.

But with the advent of defibrillation and assisted-ventilation technologies, these standards could no longer be automatically applied. In an increasing number of cases, people whose hearts were not beating, or who were not breathing, could be resuscitated and maintained well past the point where previously they would have been declared dead. With dissemination of these technologies into the community, through citizen education about CPR and, more importantly, provision of paramedics with proper training and equipment to deliver vital pre-hospital care, a significant number of resuscitated individuals—perhaps as many as twenty percent—now go on to a full recovery, literally backing out through death's door.

The creation of a brain-death standard was also driven by the need to define permissible conditions for removing organs for transplantation. The technologies for sustaining biological functions in persons unable to maintain a heartbeat or breathing on their own evolved largely in parallel with the technology to harvest and successfully transplant vital organs. From legal and ethical viewpoints, as well as for purely medical reasons, both transplant surgeons and hospital administrators wanted the moment of death, particularly in the case of artificially maintained patients, clearly defined. All organs begin to degenerate very rapidly upon death, and need to be removed from a dead donor as quickly as possible if they are to function in a living recipient. So even (or perhaps especially) when a patient maintained on life-support systems has previously expressed a willingness to donate organs upon death, someone has to declare the exact moment when death has occurred and the organs can be harvested. An increasingly litigious public has driven many hospitals and transplant surgeons to seek the protection of the law in borderline cases.

And last, but not least, remember that at the precise moment a person is declared dead, his or her legal and moral rights as a person cease completely. Health insurance to hospitals and physicians stops, and life insurance must be paid to beneficiaries. The precise moment of death may have important implications for a wide range of legal issues related to survivorship as spelled out in wills, and for those who inherit both the assets and the liabilities of the deceased. This may be especially critical when both a husband and a wife, each with separately defined lines of succession and inheritance, are critically injured and near death.

In 1968 a group of physicians met at Harvard University to establish new criteria for declaring a person dead on the basis of loss of control by the brain of circulation and breathing, and loss of integration by the brain of other critical bodily functions. They proposed that, in the absence of specific conditions such as drug intoxication or hypothermia, a person who is judged irreversibly comatose, is unable to breathe unaided, and has no neurological reflexes or electrical activity in the brain is dead. These criteria were discussed and refined over the next dozen years or so. In 1981, the President's Commission for the Study of Ethical Problems in Medicine and Biomedical and Behavioral Research, in collaboration with a variety of professional health organizations and the National Conference of Commissioners on Uniform State Laws, crafted the *Uniform Determination of Death Act* (UDDA). The UDDA was formulated in an attempt to coordinate the independent efforts of the various states and federal jurisdictions struggling to formulate a definition of death consonant with the new technologies. The UDDA quickly became the basis for legislation defining clinical death in all of the fifty states.

The UDDA proposed two fundamental criteria of death. First, *an individual with irreversible cessation of circulatory and respiratory function is dead.* This is just a restatement of previous standards for death, reaffirming their general validity. Second, *an individual with irreversible cessation of all functions of the entire brain, including the brainstem, is dead.* Again, complicating conditions such as shock, hypothermia, drug intoxication, and age less than one year were cited as requiring special examination to establish true death.

Although these criteria initially satisfied a majority of medical personnel, legal specialists, and most medical ethicists, some concern has been expressed in recent years about the appropriateness of the UDDA definition of brain death. The primary concern relates to the requirement for "whole brain" death. Although explicitly stated only in the second of the two definitions, this requirement is actually embedded in the first as well. Respiratory function (breathing) is regulated by the encephalon, the most primitive element of the brain, through its control of the diaphragm and chest-wall muscles. The brainstem also controls a variety of other reflex reactions mentioned earlier, such as body temperature, blood pressure, and the pupillary and gag reflexes. Without brainstem function, a person cannot breathe unaided, and this has become the most important criterion for determining brain death under UDDA guidelines. Various tests involving withdrawal of artificial respiration and observation of autonomous breathing have been defined. If a person cannot breathe unassisted after several well-monitored attempts, he or she can be declared dead, and no one would much argue with the decision. Even with a ventilator to assist breathing, such patients rarely survive more than a few weeks, because of the dependence of the heart and blood pressure on at least some brainstem function. However, in a few cases biological functions can be maintained for longer periods. In 1993, a California woman declared brain-dead was found to be in her fifth month of pregnancy; she was maintained on life-support systems for three months until the baby could be delivered by caesarian section. The baby did well; the mother was disconnected from life support at the end of the delivery.

But it is possible to breathe without mechanical aid, and to display a limited number of reflex reactions, with *only* encephalic brain function. The pumping function of the heart is largely independent of the cortical regions of the brain. If the heart survives the damage inflicted during myocardial ischemia and the resulting infarction, it can get along fairly well on its own. According to UDDA criteria, a person with intact encephalic brain function and a beating heart is not dead. Thus our heart-attack patient, if he truly is in a persistent vegetative state with intact brainstem function, is still legally alive; he cannot be declared dead by the hospital staff.

The persistent vegetative state was rare just a few decades ago, but a number of factors have intersected to produce an estimated 10,000 to 20,000 adult PVS patients at any given time in hospitals and nursing homes throughout the United States, in addition to perhaps half that many PVS children. This increase is largely a consequence of the technology and procedures that have brought our patient to his present situation; even ten years ago he would not likely have survived his ordeal. The term "persistent vegetative state" itself was not well defined, medically or legally, until just a few years ago. Doubtless some of the apparent increase in the number of PVS patients is a matter of better understanding among health professionals of just what PVS is, and thus more accurate recording of diagnoses. In the past PVS was often confused with coma; the major difference between the two states is that persons in a coma never open their eyes, whereas PVS patients show alternating patterns of sleep and apparent "awakeness." In neither case does the individual involved have any awareness of the surrounding world; the PVS patient and the comatose patient are both completely unconscious.

Truly comatose patients have a very poor prognosis; eighty-five percent die within four weeks. Those who do not regain consciousness within this time rarely survive a year. The major cause of death is lung infection; comatose patients do not have well-developed gag, swallowing, and coughing reflexes, which are important in keeping infectious agents out of the lungs. Although such patients retain some degree of brainstem function, they have lost portions of the brainstem controlling critical functions such as the sleep/wake cycle and gagging.

Patients may enter a true persistent vegetative state for a variety of reasons, but the major ones are severe head trauma and temporary loss of blood to the brain (*transient total brain ischemia*), usually following a heart attack. Patients entering PVS because of brain ischemia have only a slightly better long-term outlook than truly comatose patients. Roughly ten percent will recover from PVS within the first month or so after diagnosis, but three out of four of these will have moderate to severe neurological disorders that may shortly prove fatal. Recovery from PVS after the first month is very rare, and is almost always accompanied by severe neurological deficit. However, the persistent vegetative state itself, as the name implies, can last much longer than true coma. The average survival time for PVS patients is around three years; the longest surviving PVS patient "lived" forty-one years. Death normally results from infection of the lungs or bladder, or from cardiorespiratory failure. All medical authorities agree that persons in a persistent vegetative state have no ability to experience pain or discomfort. The fact that they appear to be awake, and sometimes display eye and limb movements, may make this hard to accept, especially for family members. But the experience of pain as a sensation absolutely cannot occur in an unconscious person, and PVS patients are absolutely unconscious.

Patients in a persistent vegetative state are the principal focus of the current debate about brain death. Many medical and legal experts feel that the presently framed requirement for whole-brain death is too restrictive, and that the functions regulated by the brainstem, while important to the biological survival of the organism, have little to do with what it means to be a live human being. "Human-ness," they would argue, has to do with a particular self, a particular persona, that makes each of us different. When we say that "so-and-so" died, we think of the way that individual moved and thought and talked and acted, not of his or her ability to swallow or narrow the pupils of the eye in response to light. We don't say an individual's *body* died. A human personality is characterized by a unique set of values exercised with a particular brand of reason and judgment that leads to a highly individualized response to the surrounding world. It is a way of seeing that world, of thinking about it and responding to it — of being in it — drawing upon a highly idiosyncratic collection of memories, of reflections on joy and on suffering. These are all functions of the cerebral cortex, as far as we know. If a person has lost all this, irreversibly and irretrievably, but can still gag and swallow, some would question seriously whether that "person" is still alive. None of the qualities we associate with being human reside in the encephalon, any more than they reside in the kidney or the liver or the spinal cord. Why then is the state of the encephalon the defining feature of human death? What if the encephalon itself were completely lost, but its functions were taken over mechanically — if such a person's higher cortical functions were still intact, would we be willing to declare that person dead? To some extent, this has happened; the pregnant "brain-dead" woman referred to earlier was kept biologically functional by supplying some of the chemicals ordinarily supplied by her damaged encephalon.[1] Yet, once the baby was delivered, these treatments were stopped and the ventilator was removed. Technically, it was argued, these treatments did not kill her, since she was already dead.

Proponents of a higher-brain definition of death thus argue that we should focus on the death of the *person* and not on the death of the *organism*. From this point of view, it ought to be permissible to define death as the irreversible loss of *higher-brain* (cortical) functions — specifically, conscious-ness and cognition — rather than insisting on loss of whole-brain function. Such distinctions are uniquely human. They do not arise in discussions of the death of single-cell organ-isms, or even of multicellular animals. Death of multicellu-lar organisms, as we have seen, arose from the need to dis-pose of excess DNA and the cells that house it. Somatic cells eventually die — *all* of them — while germ cells — *some* of them — gain immortality through passage into another being. In other animals, when all of the somatic cells are dead, we consider the organism dead, period; this is a defin-ition that we apply uniformly across all five kingdoms of liv-ing things. But human beings have introduced a new notion

---

[1] The necessity to use a term such as "biologically functional" is an indication of how the adoption of acceptable terminology can lag behind the reality it tries to describe. Brain-dead individuals (bodies?) kept from total somatic-cell death by a combination of chemical and mechanical means have no status in science, medicine, or the law. Legally, they are dead; biologically, they are un-defined.

into the biology of our own deaths—a hierarchy of somatic cells. Even with the whole-brain definition of death as it now stands, we make a distinction among somatic cells; for brain cells are, in the end, just another kind of somatic cell. We say that when the brainstem cells are dead, even if the vast majority of other cells in the body are kept alive by technology, a "person" is dead. The implication is that someone with intact brainstem function is alive. Yet in such an individual, the very cells that define a human organism as a "person" — the cells of the cortex — may be completely dead.

This redefinition of death along lines that apply uniquely to human beings is just one of a number of issues that disturb many who defend the whole-brain definition of death. They ask themselves why such a redefinition should be necessary. Death is a wide-ranging biological phenomenon, and death in all other animals is defined simply as total somatic cell death. It is unlikely that we would label as dead any other organism whose heart still beats, whose blood still flows. Why then do we need a special definition of death for humans? To save money for insurance companies and hospitals? To make it easier to harvest organs for transplantation? Death is a singular and sacred event in a human life — ought we to be redefining it for reasons of convenience or economy?

One of the criticisms of the whole-brain definition of death has been that no one is quite sure what "whole-brain" death means. The formulators of this concept obviously intended it to mean that someone cannot be declared brain-dead unless the brainstem, as well as the cortex, is dead. But how much of the brainstem must be dead to meet this criterion? Half? Three-quarters? Every single cell? How would we prove that every single cell in the encephalon is dead? In fact, it is rare that patients declared brain-dead under current guidelines have demonstrably lost every single function of the encephalon. And why stop there? The encephalon leads directly into the spinal cord. Are there no functions even one centimeter beyond the tip of the encephalon that we would equate with life? If so, what is the basis for excluding them?

But defenders of the whole-brain concept put the ball right back in their opponents' court. How much of the *cortex* would have to be dead to define the death of a human being? Half? Three-quarters? Every single cell? And how would we prove that every single cell in the cortex is dead? They argue that we simply do not know with certainty exactly which life functions are attributable to which regions of the brain. We do not really know whether or to what extent (although admittedly it would be small) some regions of the encephalon may affect things like personality or memory or a sense of humor. And we have no way of determining precisely how many cells in which regions of the cortex are dead. If we do not know absolutely that the persistently vegetative patient is *not* alive, should we not err on the side of life, however reduced its definition, however slim its possibility?

In the end many people, including many eminent bioethicists, are simply uncomfortable with declaring dead someone who retains *any* brain function. One of the more recent and eloquent iterations of this position was set forth by Dr. James Bernat, professor of neurosurgery at Dartmouth Medical School:

> ... there is a clear conceptual distinction between the hopelessly brain-damaged patient in a persistent vegetative state, and the patient who is dead. ... It is counterintuitive to the concept of death to imagine that physicians would have to observe a patient for several weeks or months before they could determine if he was dead. ... Practical conflicts arise when we consider declaring dead, patients who are in persistent vegetative states. ... Should they be buried or cremated while still in possession of motor, sensory, and autonomic behaviors? If not, should they first be given an injection of high-dose barbiturate to abolish these behaviors? Why should such an injection be necessary if they are already dead?

It may be that we as a society will never be able to define the exact and true moment of death, and perhaps we shouldn't try. First of all, it is unlikely that a pluralistic, multicultural society such as ours will ever arrive at a definition of humanness or personhood, or even life and death, acceptable to all or even to a majority of people. The highly emotional and long-standing debate about abortion indicates clearly the difficulty of defining the moment when the life of an individual begins. In the words of Robert Veatch, professor of medical ethics at Georgetown University and one of the most forceful critics of the whole-brain definition, "The determination of who is alive — who has full moral standing as a member of a human community — is fundamentally a moral, philosophical, or religious determination, not a scientific one." Similar sentiments are expressed by at least some defenders of the whole-brain definition of death, like bioethicists Jeffrey Botkin and Stephen Post: "the moment of death is not a specific physiologic event amenable to scientific determination. Rather, it is a moment defined by philosophic concepts — concepts that speak to what it means to be alive. Since such philosophic contentions defy objective proof, the moment of death must be seen as an event fixed by social consensus."

And yet we have a problem on our hands, a problem generated by technology that we as a society have created. It is a problem that we as a society must try to solve. Thousands upon thousands of patients are now suspended in PVS. Such individuals far outnumber the fortunate few whose lives are actually restored to normal by our resuscitative technologies, and their ranks increase daily. Maintaining them can sometimes place a devastating emotional burden on their loved ones, and a crushing financial burden on society as a whole. What are we to do?

This dilemma may be eased by the current trend toward giving more weight to advance directives written by the patient, or, when there are no directives or the patient is incompetent, recognizing the rights of an appropriate surrogate (usually a first-degree relative or legal guardian) to make such decisions. Advance directives (sometimes called "living wills") that clearly express the wishes of an individual not to be maintained in a persistent vegetative state are now honored in all fifty states. In fact, a federal law that went into effect in December 1991 requires hospitals to ask patients upon admission whether they have completed an advance directive, and if not, to assist them in doing so. Of course this is of no help to patients who are incompetent at the time of admission — those in a coma, for example, or individuals who are severely retarded. Most states also recognize in some form or another the right of defined surrogates to make treatment decisions on behalf of an incompetent PVS patient who has not made out an advance directive.[1]

The trend toward surrogate decision-making has evolved through a series of legal decisions made in recent years as doctors, patients, and hospitals have struggled to understand the implications and boundaries of the Uniform Determination of Death Act. This dialog between our medical and legal systems began in earnest even before the UDDA was formulated, with the landmark case of Karen Ann Quinlan. In April 1975 the twenty-one-year-old Quinlan collapsed at a party after consuming a moderate amount of alcohol while she had an alcohol-incompatible prescription sedative in her

---

[1] Although virtually all hospitals welcome and recognize living wills and durable powers of attorney identifying surrogate decision-makers, a recent study in New York and California found that a large majority of elderly persons who had made such documents failed to communicate this information to either their health-care providers or their next of kin. Anyone with health documents of this kind is urged to make their existence as widely known as possible — to the family doctor, to immediate relatives, and even to friends.

system. Shortly after she lost consciousness, her heart stopped beating and she ceased breathing; her friends administered CPR as best they could while waiting for help. Her heart and breathing stopped again after help arrived, and CPR was adminstered again, this time by a policeman. The total period of time without spontaneous respiration has been estimated at fifteen to thirty minutes. At the hospital her pulse was stabilized, but she remained unable to breathe on her own and was placed on a respirator. Spontaneous breathing returned within an hour or so but was irregular, and she was kept on a respirator that monitored her breathing pattern and provided assistance as needed. She never regained consciousness. Detailed testing over the next several months suggested she was in a persistent vegetative state. Those attending her were of the opinion that she would not recover, and this was explained to her parents, Joseph and Julia Quinlan. After a few more months with no signs of improvement, her parents asked that she be taken off the respirator. She had gradually become more dependent on mechanical assistance for breathing, and it was assumed that without it she would die peacefully in a fairly short time.

The Quinlan family's request came at a time when the debate about brain death was just beginning, and at a time when the right of surrogates to participate in decision-making was largely untested. Her ability to breathe on her own sporadically did not entirely fit the definition of whole-brain death, and the hospital and Quinlan's physicians were not sufficiently sure of the evolving guidelines to honor the family's request. Her parents then applied to the Superior Court in New Jersey, their state of residence and the state in which their daughter was being treated. The court ruled that there was no basis in law to compel a medical facility to remove a patient in her medical condition from a respirator. The parents appealed this ruling to the New Jersey Supreme Court, whose chief justice at the time was Richard Hughes, a former governor of the state.

Although about to leave for a trip to Japan, Hughes agreed (at his wife's urging) to hear the case immediately. In a 1976 opinion written almost entirely by him, the New Jersey Supreme Court issued a decision that has become the starting point for almost all subsequent law in this area. Discarding other possible bases for granting the family's request, Hughes generated a compelling case based on the constitutional guarantee of an individual's right to privacy. Hughes reasoned, as had others before him, that the right to privacy includes the right to exercise control over one's own life. The state also has a well-recognized interest in the life of each of its citizens, and may take steps to preserve and protect life that may at times interfere with the rights of the individual. But here Hughes departed from previous interpretations of the state's interest in such matters. He argued that as the power of the state — and by inference medicine and medical technology — to preserve and protect the life of an individual diminishes in the course of a natural deteriorative process, so too does the power of the state to interfere with that individual's right to control his or her own destiny. And breaking even further with precedent, he argued that when an individual is incompetent because of the deteriorative process to exercise this right of control, the right can be extended to a competent surrogate decision-maker. Moreover, Hughes advanced the notion that the appropriate surrogates in such a case should be the family, not the courts or the medical establishment.

Following this decision of the New Jersey Supreme Court, the hospital charged with Quinlan's care honored her family's request to remove her from the respirator. She was gradually weaned from mechanically assisted breathing over a period of a month or so, and then moved to a nursing home. She did not, as expected, die shortly thereafter; in fact she lived for almost another ten years, although she never regained consciousness. She continued to receive food and water through tubes. She died in June 1986 of a combination of pneumonia, endocarditis, and meningitis. An autopsy was performed with the family's permission, and her brain was preserved for further study. The results of that long and careful study were finally published, with the family's permission, in the May 1994 issue of the *New England Journal of Medicine*. It had become clear even before her death that hers was not a classic case of loss of brainstem function, since she retained the ability to breathe on her own. From the loss of all higher cognitive functions, it was presumed that she had suffered cortical damage as a result of oxygen deprivation. In fact, the most severe damage was in the area of the *thalamus* (see Figure 7 on page 108), which sorts out incoming signals and routes them to appropriate parts of the brain. The cortex was relatively undamaged. This finding was a great surprise to the medical community, and neurologists are now reassessing their models of the neuroanatomical basis of human consciousness. It reinforces the contentions of a number of bioethicists that we do not yet fully understand which life functions are associated with which regions of the brain. The pathologists' description of Karen Quinlan's brain will certainly alter the content of debates about possible higher-brain definitions of death.

The Quinlan case is the foundation on which a great deal of our thinking about a patient's "right to die" is based, although subsequent court decisions have also helped guide us in this delicate matter. The U. S. Supreme Court has declined to provide definitive guidelines, preferring to leave these matters to state courts. Even the state courts have generally been reluctant to intervene in a decision that, following the reasoning in Quinlan, is considered best resolved by the patient, or the patient's surrogate, and the doctors involved. Only when the parties simply cannot agree among themselves have the courts stepped in. Each of these cases has helped society to refine its thoughts on this issue. One case involved Claire Conroy, an eighty-four-year-old woman who suffered from brain damage essentially equivalent to PVS. She was living in a nursing home, and had been for some time completely unaware of her surroundings, with minimal brain function and no cognitive ability at all. In 1979 her only surviving relative, a nephew, was appointed her legal guardian.

In 1982, after a long series of degenerative medical episodes, Conroy became permanently dependent on a nasogastric feeding tube. The nephew felt that continued treatment was incompatible with her dignity as a human being, and asked that the feeding tube be removed — essentially, that she be allowed to die in peace.

The hospital, while fully aware of the Quinlan decision, was not at all certain that withholding food and water would be viewed in the same light as removing a patient from a ventilator. Quinlan, after all, had been maintained on just such a feeding tube for nearly ten years. The hospital refused the nephew's request, and he appealed to the courts. The original trial court agreed with the nephew and ordered the hospital to comply with his request as Conroy's legal guardian. The case was appealed, and the initial ruling was reversed by an appellate court. The nephew was preparing to take the case to the state supreme court (again in New Jersey) when Conroy died. He decided to proceed anyway, as a contribution to clarifying the law in such cases. In its 1985 decision, the court made two important points which have generally been incorporated into subsequent rulings. First, the right of a patient or a surrogate to make a decision about withdrawing *any* treatment, when that treatment no longer serves a useful end, does not necessarily have to be founded on the constitutional guarantee of privacy, but can be extracted from "the common law right to self-determination." This actually strengthens a patient's rights in such matters, because it interprets the right to die as a fundamental human right, rather than a political right potentially subject to a court's interpretation of the constitution. Second, the court ruled that artificial delivery of food or water that substitutes for a patient's ability to eat and drink is a form of medical treatment, just like a ventilator that replaces a patient's ability to breathe. Therefore, removing tubes that deliver food and water should be viewed in the same way as removing a patient from a ventilator. Patients have an inherent right to refuse such treatment, and with appropriate safeguards this right may be extended to their surrogates.

Do such decisions pave the way for courses of action that may not be in the best interests of the patient? Could they start us down the infamous "slippery slope" toward sanctioned euthanasia? Many fear that they may, and we are right to be cautious. But the record suggests that such fears may be misplaced. Take the case of Helga Wanglie, an eighty-six-year-old Minneapolis woman who had been in a persistent vegetative state requiring ventilator support for just over a year when the hospital sought the family's permission to remove her from the ventilator. The hospital felt that under the circumstances continued treatment would in no way contribute to the patient's welfare, and was therefore inappropriate. The patient's husband refused permission to withdraw treatment, and the hospital went to court, asking that an independent surrogate be appointed for Wanglie. In July 1991, the Minnesota court affirmed the husband's right to act as Wanglie's surrogate, further consolidating earlier legal opinions. But it also ruled that while it may be the duty of courts to define and protect the rights of patients and surrogates to make such decisions, courts have no jurisdiction over the content of a surrogate's decision, as long as that decision is one the patient might reasonably have made on his or her own, and unless that decision can be shown to be clearly not in the patient's best interest. The court ordered that having clearly met this test, the family's wishes must be respected. The issue of who should bear the financial cost of continued treatment was never resolved; Helga Wanglie died of septicemia a few days after the court rendered its verdict.

And so, although significant differences among the states remain, an emerging consensus has begun to shape the interaction of doctors and hospitals with patients or their surrogates. First and foremost, both the medical and legal establishments greatly prefer to see clearly written advance directives by the patient. Virtually no one will argue with these anymore. In the absence of such directives, or when the patient is incompetent to provide such information at the time of admission, first-degree relatives or legal guardians are being granted increasing latitude and power as surrogates to decide when treatment of hopelessly brain-damaged patients should be terminated. A surrogate may act on knowledge of what the patient would have wished, or failing that, may make any considered decision the patient might reasonably have made. The decision may be to continue *or* to discontinue treatment; it is the *ability* to decide that is increasingly being guaranteed; the *content* of the decision is not a challengeable issue. These types of actions are being taken with fewer and fewer references to the legal system; most of the major questions have been addressed by the courts where needed, and the medical profession seems more confident that this approach is likely to be approved by society generally. Hospitals generally review this procedure to the extent of validating the status of the surrogate, and the reasonableness of the decision, but not beyond. An impressively large number of medical experts and bioethicists are actively promoting this trend, which in effect allows the patient to choose his or her own definition of death.[1]

Unquestionably, dealing with the death of someone close is traumatic, combining an enormous sense of loss with a foreshadowing of our own mortality. In the case of a loved one who has lingered in a persistent vegetative state, the trauma may be partially offset by gradual accommodation and acceptance of the inevitable, but it can also be heightened if we have to be involved in the final decision that death should be allowed to come, even when we know in our hearts that that is what the patient would have wanted. If making this decision is complicated by having to enter into an adversarial situation with the medical establishment, the pain can only be intensified. Surrogates should not have to bear the burden of proving that their loved one meets a specified "ob-

---

[1] The President's Commission for the Study of Ethical Problems in Medicine; the Hastings Center; the American Academy of Neurology; the American Medical Association through its Council on Scientific Affairs and its Council on Ethical and Judicial Affairs; and the United Kingdom Institute of Medical Ethics, to name but a few.

jective" criterion for death. Increasingly, families and legally recognized surrogates can simply decide for themselves what should be done, with the knowledge that their right to make a moral, philosophical, or religious determination of when death has occurred will be recognized by the civil authorities.

This approach will not solve all our problems, however, and may even create a few new ones. What would happen if a shooting victim were to enter a persistent vegetative state and after a few weeks or months, when the possibility of a recovery was deemed hopeless, the family decided to withhold food and water — the usual method for bringing such cases to closure. Who would bear the responsibility for the resulting death — the attacker or the family? This defense has actually been tried in a criminal case, and was rebuffed, but we may not have heard the last of it. And who bears the financial responsibility for those choosing a whole-brain definition of death requiring expensive long-term maintenance? Insurance companies may simply write such "catastrophic health care" out of their policies; many already have. In the case of individuals with families, the family would presumably have to pick up the cost. But what about the occasional patient with an advance directive specifying strict adherence to whole-brain death and treatment at all costs, but without a family or insurance? Or the patient without a directive and without a surrogate, especially if he or she is mentally incompetent? Should the state pay the costs in these cases, for example through Medicare? Can the state be the surrogate, or appoint a surrogate?

It is conceivable that in certain limited and clearly delineated instances it may be useful to have a definition of death based not on specific anatomical considerations but, as Veatch has suggested, on the irreversible loss of consciousness. Such an understanding of death might also alleviate one of the more troubling aspects of ending care for patients in a persistent vegetative state. Societal sanctions and legal protections aside, withholding food and water can seem uncomfortably close to active killing, for both medical staff and patient families. To fail to resuscitate, or even to fail to supply crucial medicine, can be rationalized as "letting nature take its course"; both simply involve a failure to take action. Withholding food and water involves the purposeful initiation of a process that will inevitably and shortly lead to total somatic-cell death. Some, health-care professionals included, agonize over whether this is not a form of homicide, or at best, active euthanasia. But when the decision was made and accepted that patients with no brainstem function are dead, there was also an initial unease among many caregivers about removing such patients from life-support systems. But now, although this is still never done casually or unfeelingly, it has become accepted as a normal and necessary part of medical practice. If we as a society could agree that patients after some specified length of time in PVS are also clinically dead, it is possible that this final act, while still incredibly painful, might be made less traumatic for all concerned.

As resuscitative procedures continue to improve, along with the ability to maintain persons in a persistent vegetative state for long periods of time, the associated problems can only become larger and more pressing. Both Medicare and private insurance companies represent a pooling of resources by collectives of people seeking to underwrite the costs of their medical care. If maintenance in a persistent vegetative state cannot possibly result in restoration of life, yet draws large amounts of money from the common pool over indefinite periods of time, then decisions to maintain people in PVS should have the full and informed consent of all members of the pool — of society. It may thus be time to move this debate out of the academic journals and professional meetings into the larger arena of political discussion. And we should all join in that discussion.

6

Standing at the Abyss: Viruses, Spores, and the Meaning of Life

Not people die but worlds die in them.
—*Yevgeny Yevtushenko*

Even under the most sophisticated microscope, there is very little difference in the appearance of a cell at the moment of its death and one that is perfectly healthy, just as there is little in appearance to distinguish someone who has just died from someone who just fell asleep. Within a few moments a dead cell may burst open or fragment into apoptotic bodies, but if the cell is kept at room temperature it may not manifest these hallmarks of death for several hours.

So how do we know when a cell is dead? This is an important question, because the death of a living organism begins, as we have seen, with the death of a portion of its cells. If we are to understand fully the meaning of death, we need to understand what it is that is switched off inside an individual cell when it dies—what it is that returns it to chaos and to silence.

When biologists look at cells under a microscope, there are several practical means they can use to decide if a particular cell is dead or alive—the cellular equivalents of taking a pulse or checking for signs of breathing. However, these means are inexact. For example, a dye like Trypan Blue is excluded from live cells, and taken up by dead cells, making them appear bright blue. But what such dyes really measure is the ability of the cell's plasma membrane to exclude or take up the dye. Practice has shown that this correlates reasonably well with the cell's being alive or dead, but it is nothing more than a reasonable correlation. There is no way to be absolutely sure of the vital state of a cell by looking at it, unless it is so dead it has begun to fall apart. So what exactly is it that is missing in a cell when it dies? What qualities define a cell as being alive, the absence of which would make it dead?

There are many criteria used to define life in a cell. Perhaps the most important one is the ability to consume energy-rich materials (nutrients), extract the energy from them, and then use that energy to carry out the various chemical reactions supporting life within the cell. All living things do this; it is a process known as *metabolism*. Cells use energy derived in this fashion — metabolic energy — to form their structural and functional components, to reproduce themselves, either sexually or by simple fission, and to respond to the environment, for example to move about in search of food, or to escape from toxins or predators. All single-cell organisms, and multicellular organisms that do not generate their own heat, are also dependent on ambient thermal energy from the sun. The biochemical reactions necessary to extract energy from food simply do not work well as temperatures drop toward the freezing point of water, because these reactions are almost always dependent for their chemical integrity on water in the liquid rather than the solid state. "Warm-blooded" animals use metabolic energy together with ambient solar energy to keep their internal temperatures within a reasonable working range.

For most biologists, then, the various definitions of life can nearly all be traced back to the presence within the cell of an active metabolism — an ability to extract energy from food and use it to carry out the range of biological functions that we call life. But the definition of life (and indirectly the definition of death) becomes more complicated when we consider a survival mechanism called *cryptobiosis*, in which the organism using it, by virtually any criterion we could apply — including intracellular metabolism — would in many cases be dead. The dilemma this peculiar state poses is that it is reversible. At the end of the cryptobiotic state the organism reemerges, completely restored to life — able to feed, to move about, and to reproduce its own kind. This phenomenon was studied intensely in Europe in the nineteenth century, first of all because many people found it hard to believe, but secondly because it stirred popular as well as professional speculation about the possibilities and very nature of resurrection as described in the Bible. These kinds of discussions may seem sophomoric today, but issues like spontaneous generation and evolution, along with the revival of spores, posed major challenges to much of the inherited wisdom of the nineteenth century.

Cryptobiosis is one of a number of strategies cells developed to deal with an emerging crisis a billion or so years after life first appeared. As cells happily followed the dictum to be fruitful and multiply, the inevitable happened: they did, and life on earth began to get pretty crowded. Living organisms had to resort to increasingly clever tactics to stay on top in the fight for ever-diminishing resources. Cells began to push into rather unlikely biological niches to escape from the intense competition. There are certain environments on earth that can support life for most of a yearly cycle, except for periods of extreme temperature, or lack of water or nutrients, or perhaps extreme salinity. In order to exploit these niches during the seasons when conditions for life were favorable, some organisms learned to fake death during the off-season; they became cryptobiotic.

This reversible deathlike state was known already to moneran bacteria, quite likely before eukaryotic life forms ever appeared, and a number of them still practice it today.

It is one of the few ways they have to escape the only kind of death they know — accidental death. Bacteria in this state are known as *spores*. Cryptobiosis is also used by many protists, which in their cryptobiotic form are generally referred to as *cysts*. And it is even used by a few multicellular animals, as we shall see.

In bacteria, the process of spore formation, or *sporulation*, has been studied most intensively in the genus *Bacillus*. The most common signal inducing sporulation in these bacteria is depletion of nutrients from the environment. Rather than simply starving to death, these cells have devised a means of hanging around until things get better. The fall in levels of intracellular ATP (and the related molecule GTP) resulting from starvation causes the cells first of all to stop dividing and to enter into a stationary phase. This is followed shortly by the expression of a specific genetic program concerned with sporulation. The genes in this pathway guide the cell through a series of events that starts out somewhat like another round of cell division. But in this case the two daughter cells are not equal, and they do not pull apart. One daughter begins the transition into a spore; the other daughter wraps itself around the spore-daughter and helps it in its transition. One of the first things that must be done is to surround the developing spore with a thick protective coat that will prevent damage by chemicals. Unlike a fully viable bacterium, the spore will be unable to repair damage to itself. This introduces a serious problem: externally induced damage may accumulate beyond the point where the spore can function once it is revived. It is therefore essential that potentially harmful substances be kept out.

Both daughters contribute energy and materials to production of the spore coat. Once the coat is finished, water is drained from the spore and its important role in stabilizing internal cell structures is often replaced with a simple sugar called *trehalose*. As water disappears, the shrinking cell is pumped full of calcium and stimulated to produce a compound called *dipicolinic acid* (DPA). These two molecules complex with protein structures in the sporulating cell to make it rigid, and at the same time highly resistant to heat and radiation damage. When all this is finished, the nurturing daughter dies, disintegrating and releasing the mature spore-daughter into the surrounding environment.

What is released is a dry, hollowed-out shell of the original cell. It is like a city with no people in it. The basic structure of the cell is preserved exactly as it was before. The all-important DNA is intact, although coiled tightly into a dense strand that cannot be read. Cellular organelles such as ribosomes are preserved for the anticipated return to a living state; modest stores of food may be set aside to help get things restarted. But the spore is without internal motion or metabolic activity. It does not have enough water to support metabolic processes. It does not take in nutrients, nor does it extract energy from the environment or from stored nutrients. It does not need energy; all its energy-consuming activities are shut down. That there is no requirement for metabolism in cryptobiosis is demonstrated by the fact that spores dried even further and frozen to temperatures near absolute zero, where no metabolic processes known to science could possibly take place, can recover perfectly well when returned to room temperature. We will return to this point shortly.

A spore can remain in the cryptobiotic state for many years; survival for periods of fifty or 100 years or longer, although unusual, has been convincingly demonstrated. Spores are resistant to conditions that would be lethal for a living cell—extremes of temperature or drought, intense radiation, or lack of food and water. In fact, they are among the most resistant biological structures on the face of the earth. Spores also serve to distribute the bacteria they represent over a wider area, where conditions may be more supportive of life. Almost weightless, spores can be picked up and transported long distances on even gentle breezes.

Because of their unusual stability, bacterial spores present an interesting environmental challenge to humans. Disease-causing bacteria in their cellular form are readily killed by simple treatments such as exposure to soap or other mild chemicals, or heating to modest temperatures. But spores from these same bacteria are unfazed by such procedures; the only safe and practical way to kill them is by steam at high pressure. This is the principle behind a scientific instrument called the autoclave, and its home equivalent, the pressure cooker.

Although for all practical purposes dead, a spore clearly

remains sensitive to its environment. It can tell when conditions to support life have returned to normal. Usually this means the reappearance of food. When food substances come in contact with the spore's outer coat, they trigger a series of reactions that completely reverse the process of spore formation and restore the spore to a full and normal life as a bacterium. The coat breaks down; the calcium and DPA are flushed out of the cell and replaced with water. The trehalose often acts as a nutrient to get the cell up and running. Eventually external nutrients rush in and are promptly metabolized into usable energy. After a round or two of cell division, a revived bacterium is impossible to distinguish from a bacterium that has never sporulated.

Apparently, then, a spore is not dead — but why not? If it shows absolutely no evidence of life, can it truly be considered a living thing? What property does it retain that allows us to define it as alive? Reversibility of the deathlike state is an intuitively attractive way out of the dilemma, but what exactly does that mean? We know that gradually, over time, spores fail to respond to conditions favorable to growth by reviving. Did such spores "die" during the spore period? If so, what was different about them before and after they died? What thin line did they cross? If we cannot answer such questions, we really cannot understand what death is. These questions are as difficult for biologists as they are for philosophers. But if we follow the trail of cryptobiosis a bit further, we may begin to get a hint of where the answer may lie.

Protists continued the tradition of cryptobiosis, refining it and improving on it. Protists are even more fragile than bacteria, and more sensitive to changes in the environment. Like bacteria, they enter into a cryptobiotic state in response to adverse conditions — overcrowding and a build-up of excretory products, lack of food or water, too little oxygen, too much or too little salt. One of the first things most protists do as they begin the process of *encystment* is to curl up into a sphere, to minimize their surface-volume ratio. Then they begin a process that will strip them of any unnecessary baggage as they head into a state that will mimic death. Final sets of instructions are read from the DNA; some of these instructions direct the encystment process; other messages will be stored for use when and if the encysted cell returns to active life. Then the DNA is shut down and wrapped tightly in histone proteins. If the cell has macronuclei, these are fused, and many of the excess DNA copies are destroyed.

Any remaining food in the cell, along with cell parts not absolutely needed for survival, is burned for energy to fuel the encystment process. Extra mitochondria are set up to help churn out the enormous amount of ATP that will be needed over the next few hours. Most of this energy is used to manufacture a tough outer coat that serves the same protective function as the coat around a bacterial spore. The final rounds of protein synthesis are pushed through the ribosomes, and then the ribosomes are shut down. Some of them are burned together with excess mitochondria to provide a last bolus of energy to complete the encystment process. As the coat is being assembled, water begins to be pumped out of the cell, reducing the overall cell volume by as much as ninety percent. In many cases the lost water is again replaced by trehalose, which again helps keep cell structures from losing their shape during the cryptobiotic period, and provides a modest store of food for the recovery period.

The resulting cysts, like bacterial spores, are incredibly durable. In most cases they exhibit absolutely no metabolic processes, mostly because of the almost total lack of free water. They can withstand temperatures ranging from near absolute zero to well over 100 degrees centigrade. They can survive for periods ranging from weeks to decades. Like spores, they too can sense when external conditions have returned to normal, and then the process of *excystment* reverses the cryptobiotic state in a matter of hours.

Cryptobiosis is not used much in multicellular animals. A few vertebrates have developed hybernation strategies to reduce energy demands during periods of cold or deprivation of food or water, but these states never come even close to the deathlike state of cryptobiosis. There is, however, one important exception: the embryos of the brine shrimp *Artemia salina*. These are not simple single-cell organisms; by moneran and even protist standards, they are giants. These tiny crustaceans belong to the same taxonomic group as the shrimp we use to make jambalaya. They are grown and harvested commercially for use as food on fish farms. As their name implies, brine shrimp live in waters with unusually high salt content. They are found in evaporating saltwater ponds around oceans or seas, where the salt concentration may be two to eight times that of ordinary sea water. Brine shrimp produce normal eggs and embryos that proceed to develop in an ordinary fashion into adult organisms. But they also produce, particularly as the dry season approaches, embryos whose development is arrested at an early (but definitely multicellular) stage, and that are encysted within chitinous shells similar to those encasing insects like beetles. These embryonic cysts can survive even if the salt pond evaporates to complete dryness. The cysts are almost completely without water; they are tough, dry, minute particles that might be mistaken for sand. They can exist in this state, showing absolutely no measurable indication of life, for many years, which is what makes them valuable as fish food. Each dried cyst contains enough stored yolklike nutrients to supply food to the embryo upon excystment, which occurs simply by exposing the dried embryonic cysts to water. Once excysted, unless eaten by hungry fish fry, the embryos continue on their journey to becoming full-grown brine shrimp, as if nothing at all had happened along the way.

Artemia embryo cysts would hardly be worth mentioning, scarcely more than an evolutionary curiosity, if it weren't for a simple yet profound experiment that tells us a great deal about the definition of life and death at the cellular level. In this experiment, carried out at Yale University in the early 1960s by Art Skoultchi and Harold Morowitz, a batch of desiccated *Artemia* cysts was divided in half. One half was kept at room temperature; the other half was frozen in liquid helium at reduced pressure to an extremely low temperature — less than 2.2 degrees above absolute zero (i.e., less than 2.2 degrees Kelvin) — and held there for six days. Absolute zero is -460° on the Fahrenheit scale. It is the point at which all known physical processes come to a halt. Not only can there be no *biological* activity at 2° K; the very motion of atoms themselves is brought close to a standstill, and there is no energy or momentum in the system. At -460° F, the inside of a dehydrated cyst is like outer space: frozen, inanimate matter surrounded by essentially zero energy. The implications of bringing a biological organism to this temperature were summarized by the authors:

> At these temperatures the only feature of the organism that persists is its structure. While each atom retains its position, its momentum goes to zero. . . . Warming the system is a random process, so that the momentum distribution after exposure to temperatures near absolute zero is independent of the momentum distribution before freezing. . . . If an organism survives this process, all the information required for a viable system capable of responding in the appropriate biological way must be stored in its structure.

When the frozen cysts were brought back to room temperature and placed in dilute salt water, the percentage of cysts that hatched successfully was not statistically different from that found in the cysts that had been kept at room temperature. The authors concluded the following from their experiment:

> The survival of a complex biological system such as Artemia
> cysts after treatment at temperatures near absolute zero [argues]
> that all the information necessary for the specification of a liv-
> ing system is stored in the three-dimensional configuration of
> its atoms.

In other words, at -460° F, it is impossible for any dynamic "vital principle" to exist in accordance with any known principles of physics. All that is left is a particular arrangement of molecules and atoms in three-dimensional space. At extremely low temperatures, life, by any biological criterion we might choose to define it, is missing in these cells. By restoring ambient thermal energy and replacing the water necessary to support chemical reactions inside these cells, we can restore what is, by any biological criterion we might choose, life. A very powerful argument can then be made that at the level of individual cells, *the possibility of life can be defined as the interaction of universally distributed thermodynamic energy with specified biological macromolecules arranged in specified structures.* We must say "the possibility of life," because this interaction alone will not generate life as we understand it; it simply creates molecular motions within the structure that allow cells to convert food and oxygen into biologically usable forms of energy, which in turn enable the cells to carry out their ultimate mission — to sustain and protect DNA, and to enhance the ability of DNA to reproduce itself. *But if the external conditions are right — if the food and oxygen are there — then life will proceed.*

Buried in this definition of life is the clear implication

that if we could reproduce these exact structures artificially and allow them to interact with ambient thermal energy and supply them with food, we would be able to produce a living cell — to produce life. There is absolutely no reason, based on our present understanding of biology and physics, to think that this would not happen. The structure of each of the molecules in a cell, as well as its three-dimensional relationship to other molecules, is ultimately determined by the cell's DNA. Reproducing these structures and interrelationships would certainly be very difficult, and beyond present technology, but theoretically not impossible. This is an unsettling thought in many ways. We generally consider life to be fundamentally undefinable, an unbridgeable abyss we cannot cross. We approach this abyss with understandable trepidation.

Although this definition of life at the level of individual cells is extremely persuasive, it is one that very few biologists are actually conversant with. Most biologists tend to think of life at the level of whole multicellular organisms, where the activities of millions if not billions of cells are coordinated by a central nervous system — in higher animals, a brain. Death is normally defined by the loss of this coordination, which then leads to collapse of the entire system and catastrophic death of all cells. But in fact, even in multicellular organisms death always begins with the death of single cells.

If life is the interaction of structure with energy, then it follows that death at the level of a single cell must represent the loss of either structure or energy. In the case of *Artemia* cysts, if thermal energy were never returned to them, they could never return to a living state no matter how perfectly preserved their structure might be. On the other hand, we know that in spite of their tough coating they will gradually be degraded by toxic oxygen or other environmental chemicals, or by incident radiation coming through the atmosphere. Because they are metabolically inert, they cannot repair this damage, which may accumulate until their structure is altered to a point where application of thermal energy would not restart metabolism — would not restore life. This is the likely answer to a question posed earlier about bacterial spores: what thin line has a spore crossed when it can no longer be revived? Almost certainly, some critical feature of its structure has degraded over time such that the application of energy to that structure no longer initiates those reactions that we call life. And so it is dead.

We began an earlier chapter with the question *Why death*? But that is an asymmetric question; in order to answer it, we must ask another — *Why life*? While the study of cryptobiotes points up some of the ambiguities about definitions of life and death, there is another biological entity that brings us even closer to some of the most fundamental questions about the nature of life itself and thus, indirectly, about the nature of death — viruses.

Viruses have none of the characteristics of living cells to begin with. Not only are they metabolically inert, like cryptobiotes, but they also lack any of the structural characteristics we would associate with a cellular origin, such as a nucleus, mitochondria, ribosomes, membrane pumps, and so forth. They have a coat, but there is almost nothing under it. There is no way that a virus could be considered alive, in the sense we ordinarily use that word. Yet if we allow that the single most important task of living things is to pass on their genes (DNA) to as many offspring as possible, then viruses are very much a form of life. For the one thing viruses do have under their coats is genes; they do have DNA (or RNA, which the virus, once inside a cell, can convert into DNA). In fact, viruses are nothing more than DNA (or RNA) wrapped in a few strands of protein. And by the criterion of reproductive capacity, pound for pound viruses may be the most efficient biological entities around. The fact that they must infect a living cell to reproduce should not be held against them. In getting someone else to do most of the work for them, they might well be viewed as among the most successful of all life forms.

However we view them, viruses do strip the definition of life to its barest essentials. For example, viruses raise some intriguing questions about the very act of reproduction. Humans ascribe all sorts of noble reasons to their own reproductive efforts. Having children is variously held as the highest expression of the love between a man and a woman, an expression of confidence in the future of the race, and the central experience of human life. Rarely if ever would we describe our reproductive activities in terms of some common biological imperative to pass on DNA.

Higher-order reasons for generating offspring might

even be extended to creatures other than ourselves. We can imagine an element of rational reproductive will among many of the animals in our environment — horses, say, or eagles, or cats and dogs. We readily recognize their courting and mating practices; we admire their devotion to their young. But if we were to let our minds wander to the reproductive activities of lower animals like starfish or molluscs or worms, ideas about reproductive will are apt to be replaced by images of blind, mindless mating impulses. By the time we got to something like bacteria in our considerations, we would probably be at a loss to imagine what could possibly be driving them to reproduce. Love of their offspring? Belief in the future of their race? On the other hand, we would at least view them as living things, subject to whatever reproductive imperatives are involved in the propagation of life.

But what about viruses? What could possibly drive a strand of DNA wrapped up in a handful of inert proteins to want or need to reproduce itself? Where does *that* reproductive imperative — for surely it is that — come from? What, if anything, governs it?

These questions bring to mind an entirely hypothetical but nonetheless very interesting situation I heard described many years ago. It may ultimately help us to understand, or at least appreciate, the dilemma posed by viruses. In this scenario, we are asked to imagine the following. A world-famous university neurosurgeon, renowned for his ability to remove certain malignant tumors in the deepest recesses of the brain, one day learns that he himself has exactly such a tumor, in just such a place — the brainstem. At first he panics, for he knows the tumor is fatal, and he also knows that no one else in the world can remove the tumor without very likely killing him. But then, in a brilliant creative flash, he realizes that under the right conditions — with a bit of practice and a little help from his friends — he can probably remove the tumor on his own. The brain has no pain receptors; if he can get someone to open a portion of his skull under local anesthetic, then by using appropriate combinations of mirrors and surgical instruments, he should be able to perform the difficult surgical procedure himself.

After a few weeks of practicing with lights and mirrors, and running his assistants through endless drills, he is ready to go. The operation is long and difficult. Opening the skull was more painful than he thought. As familiar with the operation as he is, in the end he can't quite watch as the whining saw cuts into his own flesh and bone. He waits impatiently as his assistants lay back the skull flap and gently push apart the various layers of the brain to expose the tumor — the "easy stuff" he knows they are qualified to do. Finally, when all is ready, he picks up the scalpels and various probes from the tray in front of him. Closing his eyes for a moment, he mentally runs through the ensuing steps, sensing in his hands and in his mind's eye the necessary reversal of hand movements he has practiced for mirror-image surgery. He opens his eyes, takes a deep breath, and begins. Two hours later, he indicates by a wave of his scalpel that he has finished. The concentration required to prise the tumor from surrounding brain tissue without harming the brain itself has exhausted him completely. When he finally drops the offending tumor into the surgical specimen pan, he almost immediately falls into a deep sleep. His assistants drain and close the wound, and stitch his skull back together. To the great relief of everyone involved in this strange enterprise, the surgeon wakes the next day, and his previous symptoms rapidly disappear. It is clear the operation was a success.

Such a story, if true, would instantly be converted into several books, a TV docudrama, and very likely a movie. Our surgeon would be feted repeatedly by adoring medical groupies, and invited to speak before the American Medical Association. The best and the brightest young surgical residents would compete fiercely to get into his neurosurgery service at the university. But in focusing on the human and medical drama of this astounding achievement, we would miss an opportunity to take a fascinating look into what we might call the ultimate abyss of biology.

Let us go back over this episode from a purely biological point of view. In the movie, it will be implied that our surgeon-hero brilliantly deduced that he had a life-threatening tumor in his brain and decided on a bold course of action to remove it. But in reality, was it not the *brain itself* that made the diagnosis and mapped out the course of action? The brain, using information supplied to it during many years of medical education and training, was able to interpret correctly certain data it had acquired about its own condition. Using the eyes and ears feeding into it, consulting its data banks and employing its own capacity to reason, it determined that some of its cells had become cancerous. The brain knew it was going to die. The initial jolt of this recognition — (What else can we say of it? It *was* a recognition.) — triggered a release of neuroendocrine hormones that caused the rest of the doctor to panic.

But then the brain realized that it had all the information it needed to rescue itself from extinction; it could, in fact, repair itself. The doctor it inhabited was equipped with an excellent set of hands and fingers that the brain could control with exquisite precision. The brain directed the rest of the doctor to set up a series of lights and mirrors so that it could use other input sensors — the eyes — to guide those hands and fingers through the required sequence of manipulations. It knew that it would have to remake certain of its connections to get the hands and fingers to perform the needed manipulations correctly, so it directed the doctor to practice while it studied the reaction and rewired itself. And when it was all done, the brain accomplished exactly what it set out to do — it saved itself from certain death.

In casting about for an explanation of this peculiar yet highly directed form of behavior, it would not be unreasonable to ask what made the brain behave in just this fashion, at just this time. *Why did it want to survive?* What does it even mean to say that a brain, in and of itself, could "want" something? But can we really doubt that all of the perceptions and reactions described above were carried out in the brain, and the brain alone? Could any other part of the body have perceived and reacted in the same way? No. Simply and finally, no.

So here we have this kilogram or so of pale, mushy tissue directing a remarkable sequence of reactions designed to save itself. Why? What drove it? In trying to answer this question, some biologists — cell biologists — would take the level of analysis down one notch. After all, they would say, a brain is composed of cells; so in the end it must be individual *cells* that have the will to survive. This would not be impossible to imagine; single-cell monerans and protists are equipped with a variety of responses to assure, or at least enhance, their own survival. For example, individual cells living on their own can detect noxious substances in their environment and move away from them or retreat into a state of cryptobiosis. They can produce substances that kill or otherwise neutralize other single cells that can kill or otherwise neutralize them. These primitive cells (which certainly do not have brains) show a definite "will" to survive, so why shouldn't brain cells? If we are, in a sense, the biological heirs of these cells, perhaps some manifestation of a will to survive in our own cells is simply part of our evolutionary inheritance.

But other biologists — let us call them molecular biologists — would want to drop to an even more fundamental level of analysis. Without doubt, every single action of a cell is directed by its DNA. Like a hologram, every cell of the body has embedded in it a DNA image of our complete biological selves. If every action of a cell is guided by its DNA and nothing else, it is hard to escape the conclusion that it wasn't really the brain that sensed approaching extinction in this little fantasy, nor even the brain's cells. It must have been DNA.

Which brings us back to viruses. Viruses have certainly been on this planet much longer than human beings. Their ability to survive and reproduce is abundantly clear. In interpreting our own lives and deaths, humans tend to spin elaborate stories about love and free will. Here we stand, complex beings filled with emotions, dealing as best we can with biological imperatives we don't always understand. We are nurtured and we grow; we love, marry, and have children, whom we in turn nurture and raise to become sentient adults. And there stands the virus, a few proteins wrapped around a single strand of DNA. This smallest of all biological entities is endowed with an incredible drive to reproduce itself. This minute speck, ten thousand or more times smaller than a bacterium, straddling the line between the living and the nonliving, can lay waste to a human being in a matter of days. In forms like the Ebola or Marburg viruses, or HIV, it has the potential to wipe out a significant portion of the human species — simply by following its destiny to reproduce itself, over and over and over again. A destiny written into a simple strand of DNA.

Is this the end of our quest to understand life and death, this double strand of four nucleic acids combined into a seemingly endless string of hieroglyphics? Embedded in this hologram are instructions specifying the composition and, ultimately, the exact placing of every molecule involved in the pure structure of a cell, the structure that interacts with ambient energy and food and oxygen to allow the cell to carry out its mission. And what is that mission? It is nothing less, and nothing more, than to facilitate copying and transmission of the DNA itself to the next generation. Our own DNA seems a bit less sure of itself than viral DNA; it makes a hundred trillion copies of itself—one set for each cell in the body—to ensure the transmission of just a few copies to the next generation. And then it directs the destruction of the other hundred trillion copies. And we die.

What drives DNA to reproduce itself? Why should it bother? Do the individual nucleic acid letters or symbols that make it up spell out a message that animates the reproductive process? Does our own DNA contain this same message? Scientists in this century have deciphered the genetic code— the language in which all genes are spelled out in our DNA. But in fact, the genes that direct the construction and operation of our bodies have been found to account for only a few percent of the DNA we carry in our cells. What is written in the rest of our DNA? When translated using the standard genetic code, it is gibberish. As mentioned earlier, we now know that the vast majority of our DNA is never used to code for the proteins needed to conduct the daily business of the cell. Yet we have continued to carry this excess DNA—at great expense in terms of having to synthesize it every time a cell divides—over millions of years of evolution. So what *is* it doing? When scientists look at DNA with algorithms that are used to analyze the structure and information content of all human languages, it is here — in the so-called nonsense DNA, rather than in the regions coding for protein — that the greatest similarity to human language is found. Is this where the message that tells DNA it must reproduce itself is buried? Could this be where concepts like language and poetry and life after death, concepts that we think arise in our minds, are ultimately written? We do not know. This is where biology tapers off into pure chemistry and the thread of our existence disappears. This is the edge of the biological abyss, ill-defined though it may be, where the energy that permeates the universe interacts with structure to produce life. It is at this interface that we see both sides of ourselves, and where human beings sometimes see the face of God.

7

Coming to Closure

*Listen carefully; be attentive and alert. Death has come to
you. It is time for you to depart this world. While you must
face this reality alone, know that you are not the only one, for
death comes to all. Do not cling to life because of sentiment,
and do not fear to go on. You do not have the power to stay.*
— *The Tibetan Book of the Dead*

And so we turn once again to our patient, the man
whose myocardial cell we watched die at the very beginning
of this story, and whose heart nearly stopped beating as a result. That he did not die within minutes of the onset of his
heart attack is a tribute to his wife's prompt intervention, her
knowledge of CPR, and the swift and effective follow-up by
the first-response and advanced life-support teams. But what

is his status now? Have these interventions prolonged his life, or have they simply prolonged his death?

A little more than six weeks has passed since he was brought to the hospital. He still lies quietly in the intensive care unit: still breathing unassisted, still warm to the touch. Occasionally he moves his head or extends a finger. He emerged from deep coma and opened his eyes after the fourth day; these eyes now blink occasionally and move, but they do not see. They do not track the nurses and technicians who constantly move about him, monitoring his vital signs. They do not see the soft colors of the room, or the gleaming surfaces of the instruments efficiently arrayed around him. They do not see his wife or his two children or his friends who bend anxiously over him, hoping for some sign of recognition, some indication that he realizes they are there. But he cannot signal to them what he does not know. He is no longer able to *know* anything; it is clear that he has no cognitive function left within his brain.

Since he opened his eyes nearly six weeks ago, he has been on something approximating a sleep-wake cycle, indicating that the ascending reticular network in his brainstem is functioning. At night his eyes close and he seems almost comatose again, but during the day they open and resume their random movements. Although he seems to swallow his own saliva occasionally, he cannot chew or swallow food in an organized way, or drink and swallow water; he is given these substances through gastric feeding tubes or intravenously, as necessary. He is completely incontinent and is diapered. He recoils from loud noises, and gags or coughs when something is put in his throat. He is bathed and examined regularly, and is put through a daily regimen of assisted limb movements to prevent muscle atrophy. He has been placed on a special mattress to prevent the development of bed sores.

How did a vital, sentient, feeling human being—a man who laughed, played with his children, skied, and read books—come to be what we see before us now? He has come to this state because the brain, for all its power to create, to organize, to direct, is an exceedingly fragile and delicate organ. Above all other organs in the body, it is dependent on an uninterrupted flow of blood to deliver precious food and oxygen. The heart, of course, is also heavily dependent on the same constant supply of blood. His recent heart attack happened because of a gradual cut-off of blood to a localized region of heart muscle. This myocardial ischemia and resulting infarction involved only a small section of cardiac muscle, but one large enough that its loss, along with the muscle lost in his earlier heart attack, caused the entire heart to stop pumping for a brief period. Had the comparable event occurred in his brain—a loss of blood to a restricted region through blockage or bursting of a local blood vessel—the result would have been a stroke. But with his most recent heart attack, a combination of interrupted blood flow, and more importantly the loss of blood oxygen through interruption of his breathing, resulted in a temporary state of total brain ischemia.

Of course, all the other cells and tissues in his body also experienced the same transient total ischemia when his heart stopped beating, and his breathing faltered, but the impact of ischemia on the brain is different. Although brain cells may seem to do very little real work — they do not contract to lift things or to pump blood, nor do they manufacture huge amounts of proteins for export — they nevertheless consume oxygen, and fuel in the form of glucose, at a rate far in excess of any other cells in the body. The major use for this oxygen and fuel is to generate energy to drive the large numbers of membrane pumps found in all brain cells. As in any other cell, these pumps are absolutely essential to keep water out and to maintain ionic gradients across the cell membrane. At rest, brain cells actively pump out sodium and calcium, as well as water, and must prevent these molecules from reentering. The brain also uses pumps selectively to build up large internal stores of potassium, which it must prevent from spontaneously leaking out.

The maintenance of these ionic gradients across the outer cell membrane is what generates an electrical potential in the nerve cells (neurons) of the brain, and allows them to pass a message along to the next cell. In response to either internal or external stimuli, a given neuron can depolarize, letting the ions briefly rush toward equilibrium across the membrane, collapsing the ionic gradients and sending an electrical current racing along the cell's surface. Some neurons send out very long fibers called *axons*, which may connect either with other nerve cells, or with a target organ such as muscle. If the neuron is connected to another neuron through that cell's receiving fiber, or *dendrite*, the electrical current jumps the gap (*synapse*) between them and depolarizes that nerve cell as well. But almost before this impulse has reached the next cell, the ion pumps located all along the neuronal membrane are already at work pushing sodium ions out and potassium ions back in. This process continues from nerve cell to nerve cell until the electrical signal reaches its target. Because nerve cells do this over and over, and must collapse and restore the ionic gradients extremely rapidly, they have many more membrane pumps than most cells. And these pumps use enormous amounts of cellular energy, or ATP.

In other cells in the body, a large portion of the glucose and other nutrients brought in with blood is usually stored in the form of glycogen or fat to serve as reserves for future energy demands. When excessive demands arise, the reserves are burned together with oxygen to produce ATP, the form of energy used to drive all cell processes. It is even possible to generate a modest amount of ATP from these reserves in the absence of oxygen, in a process called *anaerobic metabolism*. If oxygen deprivation does not go on too long, anaerobic metabolism may be sufficient to meet an ordinary body cell's needs for some time.

Again the brain is different. It burns up nearly all the fuel coming into its cells almost the moment it arrives, converting it to ATP and another chemical form of usable energy called *creatine phosphate* (CP). This requires a constant, uninterrupted flow of oxygen. Few if any of the nutrients are stored for future use. The ATP and CP in turn are consumed almost immediately to fuel each cell's enormous number of membrane pumps. Thus brain cells are always living on the edge. They never relax, never store up energy reserves that might see them through anything but the briefest period of deprivation. And at least for human cells there is no possibility of entering a quiescent state like cryptobiosis until the danger is past.

The physicians overseeing our patient's case are more than aware of the fragility of the human brain. They know that very large numbers of his brain cells have been irreversibly damaged. They are now certain that he has entered a persistent vegetative state from which he will not recover. Tests with a new technology called positron-emission tomography ("PET scan") have shown that the rate of glucose metabolism in his cortex is less than thirty percent of normal. Electroencephalographic readings also suggest greatly reduced cortical function. Radioactive tracers injected into the bloodstream indicate that blood flow to the brain is less than a third the normal rate. All these measurements have been repeated several times during the past few weeks, each time with the same result. It appears that the majority of cells in his cortex have simply died. The doctors know that such patients at autopsy often show massive shrinkage of the cerebral hemispheres, which pull back from the skull case and collapse onto the brainstem. His brainstem itself seems to be in reasonably good shape, though; it is directing all the actions his body is still engaged in.

All of this has been explained sensitively and compassionately to his wife. She has also been told there is no treatment possible; damage from loss of nerve cells is not reversible. The likelihood that he will ever recover consciousness is now judged essentially nil, and even if he did, the combined damage to his heart and brain would preclude anything remotely resembling a normal life.

His wife had prepared herself mentally for this outcome, which became increasingly likely when he did not regain consciousness within the first week after his heart attack. She also understands that he cannot be declared legally dead. The body she visits every day still retains the vast majority of its somatic functions, and while it is painfully familiar to her in its physical appearance, the man she knew as her husband is irretrievably lost. It is a situation they had both foreseen and had discussed in detail. As in so many other things in their life together, they were of a single mind about what to do in such a case. With help from an attorney, they both drew up simple living wills. The state in which they reside allows them to stipulate through such advance directives that should either of them enter a persistent vegetative state, with no hope of recovery, they would wish no life-prolonging procedures to be administered, including artificial delivery of food and water. Each promised the other that they would do everything in their power to ensure that these wishes were carried out; the durable powers of attorney they granted one another gave each the legal authority to keep the promise. Each asked only that all necessary steps be taken to prevent any pain or discomfort.

Last night the patient's wife met with her two children and their spouses, and told them she wanted to request that all further treatment be suspended, specifically, that food and water be withheld. They all agreed, sadly, that this seemed the best course of action. This morning she returned to the hospital and asked that the document indicating her husband's wishes, already on file with his hospital records, be put into effect.

Later this afternoon, her husband will be disconnected from the various monitors and warning devices that have kept watch over him since his arrival in the ICU. The i.v. drips and feeding tubes delivering food, water, and medicine to his system will be removed, and he will be transferred to a simple but comfortable room on the fourth floor. Were he suffering from total brain failure — had his brainstem been damaged to the extent that he could not breathe without a respirator — disconnection from his life-support systems would bring death in a matter of minutes. The exact course of events in this case is hard to predict. Death will come from starvation and dehydration, perhaps in a few days, possibly in a week or more. But it will come soon. The nursing staff will not abandon him. They will bathe him, and keep his mouth and eyes moist. They will not exercise him, but they will move him around on the bed. This is done more to re-assure the family than to help the patient, for he is long past the possibility of any personal awareness of pain or discomfort. But it is also, the staff would tell you, for the dignity due the memory of any human being.

Most likely he will die because dehydration will induce an electrolyte imbalance in his blood, which will cause his heart to stop beating once again, interrupting the flow of blood to his brain. Then his body will experience the total brain ischemia that weeks ago deprived him of his cortical function—his humanness—but this time it will take him through to the end. There will be no attempt to stop the process; his chart will be clearly marked DNR—Do Not Resuscitate. Once his heart stops, the consequences will be rapid and irreversible. Sensing the sudden absence of oxygen and glucose, the brain will at first increase the diameter of its internal blood vessels in an attempt to bring in more blood. But the resulting increase in fluid leakage from these vessels will cause edema, swelling of the brain that will actually crush the vessels and cut off any remaining blood flow.

A sequence of events similar to what we saw earlier in the necrotic death of his heart cells will take place, but in a much compressed time scale. Any scant reserves of fuel will immediately be converted through anaerobic metabolism into ATP, which in turn will immediately be grabbed by the membrane pumps in one last, vain attempt to maintain the cell's osmotic and electrochemical balance. The ATP and CP within the neurons will drop to dangerous levels within seconds, essentially disappearing within two to three minutes. A few minutes after that, the membrane pumps will shut down permanently for lack of fuel, and allow water and calcium ions to flow unimpeded into the cells. The water will cause the cells to swell and strain against one another; the calcium will begin to corrode the mitochondria, which will sit idle for a few minutes and then puff up and start to unravel. In the nucleus, the increased sodium and calcium levels will cause the chromosomes to clump into useless, gummy strands of DNA and protein. The ribosomes, running full blast just moments before to churn out proteins needed for the patient's brain cell function, will fall apart and drift off into the cytoplasm.

Slowly at first, but with gathering speed, the individual brain cells will detach from one another, rounding up and pulling back the axons and dendrites that had sought each other out during his lifetime of thinking and remembering. Although the cells will not burst open in quite as violent a way as his heart cells did earlier, their membranes will develop massive leaks, and they will spill their contents into the fluid that bathes the brain and extends down into the spinal cord. But this time there will be no army of white cells marching in to remove the dead, or fibroblasts to lay down scar tissue. The blood in which the white cells travel will no longer reach the brain, and at any rate the white cells themselves will be trying desperately to deal with their own sudden lack of oxygen. In a few minutes they too will burst apart under the strain of inrushing water. The patient will of course be unaware of any of this, or that his body has begun the process of unraveling itself, preparing for a return to the elements from which it came — to unconnectedness, to chaos, and to silence.

Epilogue

And every time again and again
I make my lament against destruction.
                              —Yevgeny Yevtushenko

We die because our cells die. Clearly the definition of humanness must transcend descriptions that can be derived from studying the lives of individual cells; yet it is true that when death comes for us it gathers us in cell by cell. The death of our cells, as we have seen, is not an *a priori* requirement of life. It is an evolutionary consequence of the way we reproduce ourselves, and of our multicellularity. For reasons which are difficult to discern across thousands of millions of years of evolutionary time, the decision to use sex as a means of reproduction was followed, in the evolutionary line leading to human beings, by the generation of reproductively irrelevant DNA. That irrelevant DNA, segregated off into somatic cells, has become us.

DNA has only one goal: to reproduce itself. It does this in accordance with the same physical laws — the same principles of thermodynamics — that govern all the rest of the universe. Once a reasonable number of our germ cells have been given a chance to impart their DNA to the next generation, our somatic cells become so much excess baggage. They serve no useful function, and they — *we* — must die, so that change can be transmitted to the next generation.

Under the guidance of DNA, every somatic cell in the body will senesce and ultimately die on its own. This has come to be known as *programmed death*. If cells escape accidental cell death, they will ultimately be instructed to commit suicide — to execute the sequence of events known as *programmed cell death* via apoptosis. Most often this happens when the DNA of a cell has accumulated unacceptable levels of DNA damage. But death of the body rarely if ever occurs because of the cumulative effects of sequential, one-by-one extinction of somatic cells. Autopsies of very old people (over eighty years of age) usually reveal signs of half a dozen or more serious ailments that could have been expected to lead shortly to death. Sooner or later, as cells gradually die off through apoptosis and critical organs such as the kidneys or lungs or liver begin to fail because of cell loss, the heart will stop beating. Then the death process takes place in a greatly accelerated fashion. Deprived of nutrients and oxygen carried by the blood, the remaining cells in the body will die a violent necrotic death in a matter of minutes. Brain cells will be the first to go; the rest will soon follow.

Whether cells die by necrosis or apoptosis, the key element in their demise is the loss of the carefully integrated cellular structure that allows metabolism to sustain itself. In necrotic cell death, structure is disrupted mostly by the influx of water, stretching and tearing and pulling the cell apart. In apoptosis, internal structures (aside from DNA) are not altered so much as segregated from one another as the cell disintegrates into apoptotic bodies. The individual organelles are all there, but they can no longer interact. A collection — even a complete collection — of organelles in separate apoptotic bodies is no more a cell than a collection of body organs in separate bags is a human being. The individual organ structures are still there, and may function for awhile, but the structure of the organism is lost forever. So it is with a cell dying by apoptosis, shedding parts of itself like petals from a flower, or leaves from a tree.

The order in which somatic cells die is of no particular concern to nature, although recently it has become of increasing concern to us. When the heart stops pumping, the brain cells die first, as we have seen. Other cells follow, the timing of their deaths determined by their ability to carry out anaerobic metabolism with food stores set aside for leaner times. No one seems to have kept (or at least published)

accurate records on the point, but we could very likely recover viable cells such as fibroblasts from a human being for
quite some time after complete loss of brain function and an
official declaration of death. Put into culture, these cells
could proceed to complete their Hayflick limit for many
weeks or months after the body from which they came had
died, until they finally succumbed to their own genetically
encoded deaths by apoptosis. Unless, of course, they somehow got transformed by a virus, in which case they might
carry their DNA hologram endlessly into the future like some
renegade germ cell, or like the cells taken from Henrietta
Lacks's tumor.

We have learned to intervene in the death process started by the loss of heart cells, and in some cases this intervention works very well. The formerly rapid and irreversible
slide toward death can often be halted. A defective heart can
even be replaced, either by a transplant, or possibly one day,
by a completely artificial heart; the heart, in the end, is simply a pump. Certain types of damage to the brain can be repaired, too, but once neurons die they cannot be brought
back to life, and they cannot be replaced. The technology of
resuscitation has forced us to face squarely the issue of death
as it relates to brain function; we are moving ever closer to
the notion that the death of too many cells in the cortex of
the brain signals the death of the person, if not of the body.
Yet even this redefinition may not be sufficient to address
the problems we have created for ourselves. As Peter Singer,
the eminent Australian bioethicist, has stated in his book

Rethinking Life and Death: "The patching [up] could go on and on, but it is hard to see a long and beneficial future for an ethic as paradoxical, incoherent and dependent on pretense as our conventional ethic of life and death has become. New medical techniques, decisions in landmark legal cases, and shifts of public opinion are constantly threatening to bring the whole edifice crashing down."

Nevertheless, for the present, human beings seem to have decided that cells of the brain are more important in defining life than other cells. Nature, of course, would make no such distinction; from her point of view, a brain is no more or less important than a lung or a kidney or a foot. Nature recognizes no hierarchy among somatic cells. Why do we make such a distinction if nature does not? The brain evolved to coordinate the activities of the body more effectively, to make the organism it directs better at competing for resources, for survival, and for the right to pass on a particular set of genes — a particular pattern of DNA. But somewhere along the way the human brain took a completely unprecedented turn; it acquired *mind*. This meant nothing at all to nature, except as it might promote the welfare of DNA, but it took us, as biological organisms, into distinctly nonbiological arenas apparently having little to do with survival and reproduction: poetry, for example, or pure reason or pure mathematics; art, religion and music; sitcoms and soap operas. The pressures that govern our own further evolution are no longer strictly biological; through mind we have acquired *culture*, and that, rather than a competition for resources needed to survive until breeding age, is now the dominant selective force in our reproductive success. As Richard Dawkins has pointed out, although culture exists only in our minds, it nevertheless has its own evolutionary momentum, just as genes and DNA do.

We have yet to see how or to what extent or even whether the acquisition of mind may have changed or may yet change the natural order of things. Human beings have escaped the harsh realities of natural selection, but the rest of the biological planet has not. It is not simply that we have created mental pastimes which make our passage through life more pleasant, or at least more tolerable; through mind, we have begun to alter nature, and even our biological selves, in ways never before seen in the biosphere in which we evolved. No longer subject to the early and harsh death nature reserves for the weak or the unfit, we have begun to accumulate genetic defects that would long ago have been culled by nature. Medicine in the twentieth century has already enabled us to alter the composition of the human gene pool by keeping alive through reproductive age people who, in a more natural environment, would have died of genetic disease. Medicine in the twenty-first century, through gene therapy, will extend this trend into frontiers still only vaguely perceived, and with genetic consequences that can only be guessed at. And not content to have escaped the competition for resources imposed on other living things, we have begun to alter by pollution and to deplete by consumption those resources upon which other life forms depend. This may not be without a price. Most of us have become aware that depletion of natural resources deprives us of species whose beauty or grace or physical prowess we admire and whose absence from this planet we would mourn. What we are only dimly beginning to perceive is that these species are in turn the natural habitat of a range of microbial organisms with whom they have lived for millions of years in peaceful equilibrium. Deprived of their natural hosts some of these microorganisms, out of sheer desperation, have begun to jump to humans where there is no such equilibrium, and may not be for millions of years to come. The results, as we have seen with the Ebola and AIDS viruses, can be disastrous.

Our minds have come to regard our bodies as something more than a fancy vehicle for nurturing and transmitting DNA, and have made us unwilling to let reproduction be, as it is for all other living creatures, our only imperative, our only impact on the world in which we live. We have become thinking creatures who think about a great deal more than DNA. As the story about the brain surgeon shows, the brain through mind can even think about itself. But the mind contemplating the brain — and thus itself — is a bit like looking into a mirror with another mirror behind our back; it triggers an endless array of front and rear images tapering off into infinity. So when we try to think of the universe and our own place in it, when we think about what defines us as human, or about life and death, perhaps we should retain a certain skepticism about our conclusions. We must remember that whatever notions we may hold about the importance of the brain as mind, these notions originated in the mind as brain. In any other aspect of human affairs, this would be considered a conflict of interest. It is a humbling thought, but the human mind is not the power that drives the universe. Whether we like it or not, the mind as brain is driven ultimately by DNA, this bizarre molecule that is in turn driven — mindlessly, we presume, yet somehow *desperately* — to reproduce itself.

When we complete the dying process, every single cell in our body will be dead, as nature intended. If we have done her bidding, we will have passed on our DNA, packaged in germ cells, to the next generation. That DNA may very well be standing next to our death bed in the form of a son or a daughter. The DNA in all the rest of the cells of our body — our somatic DNA — will no longer be of any use; like the DNA in the first redundant macronucleus a billion or so years ago, it will be destroyed. To paraphrase an old biological saw, a human being is just a germ cell's way of making another germ cell — as is a cockroach; as is a cabbage. This is not a very flattering way to explain ourselves to ourselves. We want so desperately to be more than just a vehicle for DNA, and at least transiently we are. Yet somatic cells will die at the end of each generation, whether they are part of an insect wing or a human brain. We may come to understand death, but we cannot change this single, simple fact: in the larger scheme of things, it matters not a whit that some of these somatic cells contain all that we hold most dear about ourselves; our ability to think, to feel, to love — to write and read these very words. In terms of the basic process of life itself, which is the transmission of DNA from one generation to the next, all of this is just so much sound and fury, signifying certainly very little, and quite possibly nothing.

Further Reading

Books and Journals

"Adult Advanced Life Support." *Journal of the American Medical Association* 268 (1992):2199.

"Adult Basic Life Support." *Journal of the American Medical Association* 268 (1992):2184.

Bell, G. *Sex and Death in Protozoa*. Cambridge University Press: Cambridge, 1988.

Crowe, J. and J. Clegg. *Anhydrobiosis*. Dowden, Hutchinson and Ross, Inc.: Stroudsburg, PA, 1973.

Dawkins, Richard. *The Selfish Gene*. Oxford University Press: Oxford, 1989.

Defining Death: Medical, Legal and Ethical Issues in the Definition of Death. U.S. Government Printing Office: Washington, DC, 1981.

"A Definition of Irreversible Coma." (Report of the Ad Hoc Committee of the Harvard Medical School to examine the definition of brain death.) *Journal of the American Medical Association* 205 (1968): 337–340.

Gall, J. (ed.). *The Molecular Biology of Ciliated Protozoa*. Academic Press: New York, 1986.

Gold, Michael. *A Conspiracy of Cells: One Woman's Immortal Legacy and the Medical Scandal It Caused*. State University of New York Press: Albany, 1986.

Grenvik, Ake and Peter Safar, eds. *Brain Failure and Resuscitation*. Churchill Livingstone: New York, 1981.

"Guidelines for the Determination of Death." (Report of the Medical Consultants on the Diagnosis of Death, to the President's Commision for the Study of Ethical Problems in Medicine and Biomedical and Biobehavioral Research.) *Journal of the American Medical Association* 246 (1981): 2184–2186.

Halvorson, H. and A. Monroy. (eds.) *The Origin and Evolution of Sex*. Alan R. Liss, Inc.: New York, 1985.

Hayflick, Leonard. *How and Why We Age*. Ballantine Books: New York, 1994.

Kallman, F. J. "Twin data on the genetics of aging." In G.E. Wolstenholme and C. O'Connor. (eds.) *Methodology of the Study of Aging*. Little, Brown Co.: Boston, 1957.

Lamb, David. *Death, Brain Death, and Ethics*. State University of New York Press: 1985.

Margulis, Lynn and Dorion Sagan. *Origins of Sex*. Yale University Press: New Haven, 1986.

"Medical Aspects of the Persistent Vegetative State." (The Multi-Society Task Force on the Persistent Vegetative State (American Academy of Neurology).) *New England Journal of Medicine* 330 (1994): 1499, and 330: 1572.

"Persistent Vegetative State and the Decison to Withdraw or Withold Life Support." (Council on Scientific Affairs and Council on Ethical and Judicial Affairs of the American Medical Association.) *Journal of the American Medical Association* 263 (1990): 426.

Potten, C. S. (ed.) *Perspectives on Mammalian Cell Death.* Oxford University Press: Oxford, 1987.

Singer, P. *Rethinking Life and Death.* St. Martin's Press: New York, 1994.

Smith, I. R. Slepecky and P. Setlow. (eds.) *Regulation of Procaryotic Development.* American Society for Microbiology: Washington DC, 1989.

Walker, A. Earl. *Cerebral Death.* Urban and Schwartzenberg: Baltimore, 1985.

Zaner, Richard M. (ed.) *Death: Beyond Whole-Brain Criteria.* Kluwer Academic Publishers: Dordrecht, 1988.

Walton, Douglas N. *Brain Death.* Purdue University: Purdue, 1980.

Zwillig, R. and C. Balduini. (eds.) *Biology of Aging.* Springer-Verlag: Berlin, 1992.

Original Scientific Papers

Allsopp, R., et al. "Telomere length predicts replicative capacity of human fibroblasts." *Proceedings of the National Academy of Sciences* 89 (1992): 10114.

Bayreuther, K., P. Francz, J. Gogol, and K. Kontermann. "Terminal differentiation, aging, apoptosis and spontaneous transformation in fibroblast stem cell systems in vivo and in vitro." *Annals New York Academy of Science* 663 (1990): 167.

Bernat, James. "How much of the brain must die in brain death?" *Journal of Clinical Ethics* 3 (1992): 21.

Botkin, Jeffrey, and Stephen Post. "Confusion in the determination of death: distinguishing philosophy from physiology." *Perspectives in Biology and Medicine* 36 (1992): 1.

Clegg, James S. "Interrelationships between water and metabolism in

Artemia salina cysts: hydration-dehydration from the liquid and vapour phases." *J. Exp. Biology* 61 (1974): 291.

Davis, M. J. Ward, G. Herrick, and C. Allis. "Programmed cell death: apoptotic-like degradation of specific nuclei in conjugating Tetrahymena." *Developmental Biology* 154 (1992): 419.

DeBusk, F. "The Hutchison-Gilford progeria syndrome." *Journal of Pediatrics* 80 (1972): 697.

Denis, H. and J. LaCroix. "The dichotomy between germ line and somatic line, and the origin of cell mortality." *Trends in Genetics* 9 (1993): 7.

Eiichirou, U., et al. "Preserved spinal dorsal horn potentials in a brain-dead patient with Lazarus' sign: a case report." *Journal of Neuro-surgery* 76 (1992): 710.

Gillar, P. J., et al. "Progessive early dermatologic changes in Hutchison-Gilford progeria syndrome." *Pediatric Dermatology* 8 (1991): 199.

Hastie, N., M. Dempster, M. Dunlop, A. Thompson, D. Green, and R. Allshire. "Telomere reduction in human colorectal carcinoma and with aging." *Nature* 346 (1990): 866.

Hayflick, Leonard. "The cell biology of human aging." *Scientific American* 242 (1980): 58.

Kalimo, H., et al. "The ultrastructure of brain death." *Virchow's Arch. B Cell Pathol.* 25 (1977): 207.

Greider, C. and E. Blackburn. "Telomeres, telomerase and cancer." *Scientific Americani* 274 (1996): 92.

Keilin, D. "The problem of anabiosis or latent life: history and current concept." *Royal Society of London, Proceedings.* Series B, 150 (1959): 149.

Kerr, J., A. Wyllie, and A. Currie. "Apoptosis: a basic biological phenomenon with wide-ranging implications in tissue kinetics." *Brit. J. Cancer* 26 (1972): 239.

Mantegna, R. N., et al. "Linguistic features of noncoding DNA sequences." *Physical Review Letters* 73 (1994): 3169.

Oppenheim, R. "Cell death during development of the nervous system." *Annual Review of Neurosciences* 14 (1991): 453.

Örstan, A. "How to define life: A hierarchical approach." *Perspectives in Biology and Medicine* 33 (1990): 391.

Rosenberger, R. "The initiation of senescence and its relationship to embryonic cell differentiation." *Bioessays* 17 (1995): 257.

Saunders, J. "Death in embryonic systems." *Science* 154 (1966): 604.

Skoultchi, A. and H. Morowitz. "Information storage and survival of biological systems at temperatures near absolute zero." *Yale Journal of Biology and Medicine* 37 (1964): 158.

Smith-Sonneborn, J., M. Klass, and D. Cotton. "Parental age and life span versus progeny life span in Paramecium." *J. Cell Science* 14 (1974): 691.

Yarmolinsky, M. "Programmed cell death in bacterial populations." *Science* 267 (1995): 836.

Index

ABOUT THE TYPEFACE

The typeface used in this book is a digitized version of one of the many faces descending from the influential French type cutter Claude Garamond (1499–1561). Garamond had been apprenticed to the great Parisian printer Antoine Augerau, but became one of the first to establish himself as an independent designer and cutter of type, which he supplied to others. Garamond's Roman and italic typefaces were widely imitated, and became the standard in printing for 300 years.